Rural Development in Taiwan and Mainland China

Rural Development in Taiwan and Mainland China

EDITED BY

Peter Calkins,
Wen S. Chern, and
Francis C. Tuan

Foreword by Walter P. Falcon

Routledge
Taylor & Francis Group
LONDON AND NEW YORK

First published 1992 by Westview Press, Inc.

Published 2019 by Routledge
52 Vanderbilt Avenue, New York, NY 10017
2 Park Square, Milton Park, Abingdon, Oxon OX14 4RN

Routledge is an imprint of the Taylor & Francis Group, an informa business

Library of Congress Cataloging-in-Publication Data
Rural development in Taiwan and mainland China / edited by Peter Calkins, Wen S. Chern, and Francis C. Tuan.
p. cm.—(Westview special studies in social, political, and economic development)
Includes bibliographical references and index.
ISBN 0-8133-8263-7
1. Rural development—China. 2. Rural development—Taiwan. 3. Agriculture—Economic aspects—China. 4. Agriculture—Economic aspects—Taiwan. 5. Agriculture and state—China. 6. Agriculture and state—Taiwan. 7. Produce trade—China. I. Calkins, Peter H. II. Chern, Wen S. III. Tuan, Francis C. IV. Series.
HN740.Z9C6312 1992
307.1'412'0951—dc20 **90-25624**
CIP

ISBN 13: 978-0-367-28628-6 (hbk)

Contents

List of Tables and Illustrations vii
Acknowledgments ix
Foreword, Walter P. Falcon xi

PART ONE
OVERVIEW AND HISTORICAL PERSPECTIVE

1. Overview 1
Peter Calkins, Wen S. Chern, and Francis C. Tuan

2. Chinese Rural Development: 5
Historical Parallels and Recent Hypotheses
Peter Calkins

3. Economic vs. Noneconomic Factors 25
in Chinese Rural Development
D. Gale Johnson

PART TWO
INSTITUTIONS

Introduction to Part Two 43
Francis C. Tuan

4. Technological Choice in Chinese Agriculture 47
Justin Yifu Lin
Discussion: *Wen-Fu Hsu*

5. Institutional Organizations in Taiwan's 67
Agricultural Development
Yu-Kang Mao
Discussion: *Liu Zhicheng*

PART THREE
PRICES, COSTS, AND INCOMES

Introduction to Part Three 93
Peter Calkins

6. Prices, Income, and Farm Policy in Taiwan 99
Tso-Kwei Peng
Discussion: *Guangsen Wang*

7. Prices, Costs, and Farm Income in Mainland China: Rules of Farmer Behavior in a Uniquely Mixed Economy 125
An Xi-Ji
Discussion: *Carson Kung-Hsien Wu*

PART FOUR
TRADE AND COMPETITION

Introduction to Part Four 145
Wen S. Chern

8. China's Agricultural Trade and Prospects in Foreign Competition 149
Niu Ruofeng
Discussion: *Shin-Hsing Yeh*

9. Taiwan's Agricultural Trade and Prospects in Foreign Competition 163
Hsi-huang Chen
Discussion: *Zhu Daohua*

PART FIVE
GRAIN AND RURAL DEVELOPMENT

10. The Growing Grain Deficit and Its Implications for China's Developmental Strategy 187
Dwight H. Perkins

About the Editors and Contributors 203
Index 211

Tables and Illustrations

Tables

4.1	Land and Labor Endowments in Each Province	48
4.2	Indices of Labor, Land, Tractors, Fertilizer, and Institutions	53
5.1	Changes in the Economic Structure of Taiwan, 1953-1987	69
5.2	Changes in Taiwan's Agriculture, 1953-1987	70
5.3	Tendency of Changes in Agricultural Production	88
6.1	The Relative Price of Agricultural Products in Taiwan	101
6.2	Overall Framework of Taiwan's Agricultural Policies	102
6.3	Growth Rate of Food Demand and Agricultural Production	105
6.4	The Annual Growth Rate of Agricultural Output	109
6.5	Allen's Partial Elasticities of Substitution of Input Demand	110
6.6	The Multiple Cropping Index and Diversification Index in Taiwan	114
6.7	Sources and Relative Farm Family Income	114
6.8	The Inequality of Farm Family Income	117
1	Agricultural Trade Balance in Mainland China and Taiwan	146
8.1	Imports and Exports of Major Farm Products in Mainland China	151
8.2	Changes in Mainland China's Trade Structure Between 1952 and 1988	152
8.3	Per Capita Consumption of Mainland China's Major Items of Food and Clothing	152
9.1	Indicators of Comparative Advantage	170
9.2	Nominal Trade-Weighted Foreign Exchange Rate Coefficient for Agricultural Imports, 1983/84 to 1987/88	177

9.3	Real Trade-Weighted Foreign Exchange Rate Coefficient for Agricultural Imports, 1983/84 to 1987/88	177
9.4	Nominal Trade-Weighted Foreign Exchange Rate Coefficient for Agricultural Exports, 1983/84 to 1987/88	178
9.5	Real Trade-Weighted Foreign Exchange Rate Coefficient for Agricultural Exports, 1983/84 to 1987/88	178
10.1	Grain Output in China	190
10.2	Grain Imports and Exports	191

Figures

6.1	The Trend of Relative Prices for Taiwan's Agriculture Products	101
6.2	Per Capita Consumption of Grain Relative to Income, Taiwan	104
6.3	Food Consumption in Relation to Income Level, Taiwan	104
7.1	Price, Cost, and Farm Income	129
7.2	Changes in Input Prices	131
7.3	Combination of Two Products X and Y	131
7.4	Land Use Under Household Responsibility System	133
7.5	Optimum Amount of Input With and Without Subsidized Input	133

Photographs

Free markets were reestablished to stimulate specialization	xiii
Vegetable marketing in Shenyang, Liaoning, China	41
Newly harvested corn at a farm in China	91
Cut flowers are a crop that is booming in Taiwan	143
Temporary grain storage bins at a mill in China	185

Acknowledgments

This book reports the proceedings of the symposium, "Chinese Rural Development: Strategies and Experience," organized by the China Committee of the American Agricultural Economics Association (AAEA) and held in Baton Rouge, Louisiana, on July 30, 1989. The editors would like to thank members of the committee, listed below, for their time and efforts in planning, organizing, and hosting the symposium.

The AAEA China Committee, 1989

Peter Calkins, Université Laval, Chair
Walter J. Armbruster, Farm Foundation
Colin Andre Carter, University of California-Davis
Wen S. Chern, The Ohio State University
Jong-Ying Lee, University of Florida
Pingsun Leung, University of Hawaii
Judith A. Maxwell, The Ohio State University
Francis Tuan, Economics Research Service/USDA
Gail Updegraff, Soil Conservation Service/USDA
John A. Miranowski, AAEA Board Representative

We would like also to acknowledge the funding support for the symposium from the following sponsors:

The Ford Foundation
The Rockefeller Foundation
The Farm Foundation
The Economic Research Service of the USDA

In addition, we are grateful to the Soil Conservation Service of the U.S. Department of Agriculture for supplying all materials distributed at the symposium. The Department of Agricultural Economics and Rural Sociology of The Ohio State University provided secretarial support for the preparation of this volume; in particular, we are grateful to Karen Denney for her excellent secretarial services.

We wish to extend our appreciation to Jean Johnson for editing the manuscript. We also thank Amy Boye for formatting the document for publication, as well as Linda Roberts for her helpful assistance.

Foreword

I am honored to have been asked to contribute the foreword to *Rural Development in Taiwan and Mainland China.* The occasion that led to the publication of these papers was a historic event, and the volume itself contains a number of essays of historical importance.

The significance of the 1989 Baton Rouge meeting on Chinese rural development transcends this volume, important as these papers are. The widespread agricultural successes in both Mainland China and Taiwan have been of great interest to agricultural economists for many years. Yet this was one of the first occasions when professional economists from both regions had the opportunity to meet formally and informally to exchange views on their experiences. Further excitement and concerns were generated by the Tienanmen Square incident that had occurred only two months before the session. There were early fears that the session might have to be canceled, but (most of) the show did go on. Those broader events in the Mainland served to heighten further the intensity of the conversations, especially in the corridors.

Given the rich agricultural heritage of the Chinese people in all provinces, there was much ground to cover. The general approach taken in Baton Rouge approximated a three-legged stool—presenting varying views from Taiwan, the Mainland, and several North American institutions. The discussions drew together several generations of Chinese scholars and were marked by lively debate.

Readers of this volume will find numerous issues addressed in a comparative framework: agricultural prices, land-use planning, farm size, the role of trade, agricultural organizations, human capital, and rural poverty. I am particularly impressed by the great technological achievements on crop research in both Taiwan and the Mainland, accomplished in very different institutional settings. One contrast, the much greater focus given to education and human-capital investments in Taiwan, raises interesting questions about future technical advances in Mainland China. As an Asianist (but not a Chinese scholar), I find particularly intriguing the concerns about farm size, especially of operating units, and about the working or nonworking of land markets. The importance of strong farm organizations is another common theme; but I suspect everyone will want to know more than is debated in this volume about the peer pressures and coercion that both governments used to create compliance.

The major success of this volume, I believe, is in bringing together the architects and analysts of both regions to draw useful lessons for themselves and for the rest of the world. At one level, the volume raises a fundamental question of post-1949 China, especially for the period 1949-1978. By the year 2025, for example, will these three earlier decades be viewed as a fundamental era in which Mainland China and Taiwan set off in overwhelmingly different growth patterns, determined importantly by their differing rural development strategies? More specifically, will the differences—those occasioned by regulatory policy versus price policy, a closed versus an open economy, small versus large investments in human capital, and a host of other differences—cause Mainland China and Taiwan to choose development paths that are "forever" divergent? Or, will these 30 years be seen only as a temporary deviation of paths, followed by a fairly rapid convergence of strategies and policies? My guess, given world events of the past three years (and not withstanding the Tienanmen Square incident), is on convergence. But whatever the future, the industriousness, creativity, and village organizations of Chinese people in both regions are extremely impressive and provide for lively reading.

Professor T. W. Schultz, one of the greatest contributors to the literature on rural development, said in his Nobel lecture: "Most of the people in the world are poor, so if we know the economics of being poor, we would know much of the economy that really matters. Most of the world's poor earn their living from agriculture, so if we know the economics of agriculture, we would know much of the economics of being poor."

By reasons both of size and example, what happens in China is part of the economics that truly matter. The lessons on transforming agrarian societies and alleviating rural poverty from the dual perspectives of Taiwan and Mainland China are what make the Baton Rouge meeting and this volume so important.

Walter P. Falcon
Professor and Director
Food Research Institute
Stanford University

PART ONE

Overview and Historical Perspective

Free markets were reestablished on the mainland in 1978 to stimulate specialization and market surpluses. *(Courtesy Peter Calkins)*

1

Overview

Peter Calkins, Wen S. Chern, and Francis C. Tuan

This book synthesizes the papers and discussions presented during a symposium titled "Chinese Rural Development: Strategies and Experience," held at Louisiana State University, Baton Rouge, on July 30, 1989. This symposium was held prior to the Annual Meeting of the American Agricultural Economics Association (AAEA) and was part of the program organized by the AAEA Committee for Professional Relations with the People's Republic of China (China Committee).

The symposium was a historic event. For the first time since 1949, scholars from both sides of the Taiwan Strait met to exchange information about a wide range of subjects, including agricultural organization and institutions, the role of markets, and international trade. Non-Chinese experts on the Chinese economy, such as Professors D. Gale Johnson of the University of Chicago and Dwight Perkins of Harvard University, also attended and provided important background and conceptual information.

Background Information

Since 1979, Mainland China has undertaken economic reforms and has pursued an open-door policy. As a result, there have been more opportunities for exchanges between Chinese scholars and their counterparts in Western nations, particularly the United States. During this period, a significant event happened in Taiwan. In 1987, the government in Taiwan lifted its long imposed Martial Law. Consequently, hostilities and tensions across the Taiwan Strait softened considerably, and for the first time in nearly 40 years, intellectual contacts between scholars in Taiwan and Mainland China became permissible. In view of these developments, the

AAEA China Committee decided to organize this symposium to facilitate the exchanges of scholars in agricultural economics.

The China Committee selected three broad areas to address rural developmental issues in Taiwan and Mainland China: (1) institutional organization; (2) agricultural prices, costs, and income; and (3) trade and competition. The committee invited two scholars from each side of the Taiwan Strait to write a paper on each topic. To facilitate the interactions, the committee also invited three more scholars from each side to be discussants, focusing their comments on the papers written by scholars from the other side. Therefore, the chapters in Parts Two through Five may be used for comparative analysis.

Plans for the symposium were progressing well, with all funding secured from several foundations, when the Tienanmen Square incident occurred in June 1989. There ensued a period of considerable uncertainty about the participation of scholars from Mainland China; but, after much consultation and communication, the committee decided to hold the symposium as planned. All invited scholars were able to take part in the symposium, with the exception of Dr. Justin Yifu Lin, whose paper was delivered by Mr. Chen Zhigang of Université Laval.

The dialogue portrayed in this book is not only among Chinese rural development economists, but also between two great ideological traditions in agricultural economics. The Mainland agricultural economists whose papers appear in these pages were trained in the United States before 1949 or in the Soviet Union after 1949, and have confronted widely oscillating research mandates (one exception is Dr. Lin, who earned his doctorate in the United States in the mid-1980s). On the other hand, the Taiwanese authors were generally trained in the United States after 1949. This volume thus contributes to the growing intellectual rapprochement between East and West in an age of *perestroika*.

This book also answers some fundamental questions about economic development. For example, the Taiwanese economy has had one of the fastest rates of economic development in the 20th century. To what extent is this due to wise economic strategies and policies, and to what extent to the luck of historical circumstance and natural endowments? To what degree is the increasing use of markets, free trade areas, and personal incentives in Mainland China a "rapprochement" to the Taiwanese strategy? What are the often neglected but real achievements the mainland has wrought, especially in the 1950s and the 1980s, while following a strategy of unprecedented social experimentation? And what implications do the tragic events of June 1989 have for Chinese rural development? These and other questions structure the debate that unfolds in the following chapters of the book.

During the symposium, several questions were raised concerning the impact of the Tienanmen Square incident on the roles of Chinese students currently being trained in U.S. institutions as future rural developers of China. Several speakers from Mainland China responded by expressing their confidence in the Chinese government's continuation of economic reforms and open-door policies, including the household responsibility system, land reform, a mixed planned-and-market economy, and expansion of agricultural trade. Even though these speakers were hardly in a position to enunciate government policy, they remained optimistic about the important role and potential contribution to rural development in Mainland China by U.S.-trained Chinese scholars in agricultural economics.

2

Chinese Rural Development: Historical Parallels and Recent Hypotheses[1]

Peter Calkins

Historical Parallels

Since 1949, Western observers have taken a comparative stance in viewing the economic development of mainland China and Taiwan. Explicit or implicit in their analyses was the assumption that Chinese communism was fundamentally different from Chinese capitalism and that every stride or setback was directly attributable to the disembodied political ethos nominally espoused by each regime. The propaganda of the Chinese themselves, laden with references to "running dogs of the capitalist imperialists" and "communist bandits," reinforced this perspective. Western observers all too often visited only one or the other part of China, where they were led to believe that each was the ideal political system for the modern Chinese people and a secure repository for the best in China's past. Economic apologists for each camp, for example, have written on the miracle of the mainland (or Taiwanese) model of agricultural development and mused on whether or not it could be exported.

The astounding economic development of both portions of China may spring far less from a political label than from the industriousness, cooperative spirit, willingness to pool resources, thriftiness, village and clan structure, and creativity of the Chinese people themselves. It is questionable, for example, whether the strategies of either regime could have brought about significant progress in South Asia.

Comparison of the developmental experience of the agricultural sectors of the mainland and Taiwan provides a unique social laboratory. In addition to a common cultural heritage—itself the fundamental reason why

Mao agrarianized the Marxist revolution—both regions were devastated by World War II. Both had a capital and technical infusion from a superpower hegemony: in the case of the mainland provinces, the USSR; in the case of Taiwan, the United States. Both had programs for accelerated rural development featuring land reforms in the early 1950s, an emphasis on mechanization, programs to equalize incomes, a goal of increasing irrigated area, and programs for collectivization that built upon the mutual aid heritage of the Chinese people. With food-grain self-sufficiency as a strategic goal, both tended to manage the production and trade of rice and other staple grains through cooperative organizations encouraged and closely supervised by the State, confining the non-rice sector (vegetables, fish, and pork) to private plots. Real income of the rural sectors in both parts of China fell behind that of the urban sector by almost equal percentages. Both were successful in accomplishing rapid economic development. And, at least between 1978 and June 1989, a gradual convergence in the use of input and output markets seemed to be occurring.

Preconditions and Policy

This is not to say that the two areas were alike in all respects. Three preconditions differentiated the two regimes in 1949 and contributed to a major divergence in the choice of development policy. The first and most obvious was the size and geography of the areas controlled. Taiwan's climate corresponds only to that in southeastern China, where abundant summer monsoons and warm temperatures allow for two or more crops per year. This contrasts with the area of the mainland north of the Huai River, the "northern-wheat" area, where there is a shorter growing season, much less rainfall, and greater cold; and it contrasts even more dramatically with the increasingly arid western two-thirds of the mainland.

Although both Taiwan and the mainland are mountainous, the former has 25 percent arable land[2] versus only 11 percent in the latter. Moreover, Taiwan, thanks in part to the Japanese occupation, has 69 percent of her arable land well irrigated; while only less than half of China's cultivated land is irrigated, and half of that not very effectively.[3] Irrigation has been the key to intensive multiple cropping on Taiwan and a major development goal of the mainland.[4] However, Taiwan's advantage is by no means clear. Largely because of the immigration of Nationalists, the population of the island jumped from 6.1 to 9.0 million in the period 1946-53,[5] so that as of 1975 there was only .06 ha of arable land per head of total population (.16 ha per farm resident) on Taiwan versus .16 ha (and .20 ha) on the mainland.[6]

A second precondition was the magnitude and type of assistance donated by the ideological mentor. The United States contributed $1.47 billion to Taiwan, an average of 6 percent of the Gross National Product and 30 percent of the capital invested between 1951 and 1965. Nearly 60 percent of this aid went toward developing the agricultural sector and its supporting infrastructure, and there were relatively few foreign advisors.[7] By contrast, the major thrust of Soviet aid to the mainland was human rather than monetary. At least 10,800 foreign advisors worked on the mainland between 1949 and 1958 to complete approximately 278 industrial projects, "the backbone of China's industrialization process."[8] Only $429 million in cash value was transferred, and the impact on the agricultural sector was secondary at best.

The third major precondition was that Japan had built up Taiwan only during the period when she was trying to incapacitate the mainland. Under the Japanese occupation, Taiwan was transformed from an "opium smoking traditional society"[9] to a modern one. Even though the Taiwanese were exploited and discriminated against, self-interest led the Japanese to educate them (between 1917 and 1942, the percentage of school-age children attending school rose from 13 to 66). The Japanese put in irrigation systems; built up roads, railroads, and communication systems; developed modern business institutions and commercial centers; and generally provided an excellent infrastructure for development. Between 1905 and 1944, irrigated hectares increased from 200,000 to 520,000;[10] and between 1911 and 1940, Net Domestic Product (NDP) increased two percentage points per year faster than population.

These preconditions are important in that they led to the need for differing development policies. The difference lay not in the overt divergence between communism and capitalism; rather—because of geography, aid and infrastructure—Taiwan chose to follow a consistent, natural transformation from an agricultural to an industrial economy, with any vagaries due to agricultural weather cycles; while the Communists elected to jolt the bigger mainland into motion by artificially creating policy gyrations between the traditional values Taiwan pursued and periodic reinfusions of Marxist ideology. Given the natural tendency for weather cycles to wreak havoc on agricultural planning, Mao's decision to introduce further uncertainty into production decisions in the form of ever-shifting policies was all the bolder. (Between 1949 and 1966, 17 droughts damaged over 200 million ha, and ten floods affected over 50 million ha.[11]) But the very magnitude of the task called for a constant refocusing on the "new" social forms.

It also called for exploiting a wider case of *dramatis personae*. On the mainland, the party, the army, the bureaucracy, and to some extent the mass organizations, became distinct actors called into the spotlight at critical turns during the first 30 years of development and, most recently in 1989. In Taiwan, by contrast, the army was needed mainly to help with harvests, not to control political excesses. The party and bureaucracy of Taiwan were often synonymous and exhibited no polarization of views or roles.

In the two parts of China, only the role of the mass organizations was similar: to involve the populace in the *nouveau regime* and standardize, if not mold, values. For example, on the mainland in 1953, rural supply and marketing cooperatives were the largest mass organization.[12] Cooperatives of this very type were also the backbone of the "miracle" of Taiwan: farmers were organized first into Irrigation Commissions and Farmers Associations and subsequently into Specialized Production Areas, whose cooperatives handled the schedule of planting, the joint ownership of machines, and group marketing.

Policies Required for the Mainland

In 1949, mainland China's cities were laid waste, there were inequalities in the distribution of productive assets,[13] communications networks were shattered, there was little national confidence, and corruption and venality were rampant, particularly in the cities. Strong policies were needed.

Stages of development fell into a clear alternating pattern. Type I emphasized traditional Chinese inputs, unlimited births, and the creative use of logical contradictions (thesis/antithesis)—notably, moving forward on both the industrial and agricultural "legs" so that one development concept "became two." Type II emphasized modern inputs, family planning, and synthesis of previous contradictions—notably, allowing technology and free enterprise to move the economy ahead as one, not two, sectors, i.e., on "one leg."

Type I	**Type II**
1949-54, 1958-60, 1966-9, 1975-6	1955-7, 1961-5, 1970-4, 1976-89
traditional inputs	modern inputs
pro-natal	family planning
thesis/antithesis	synthesis

Within this general alternation, each stage embodied specific short-term steps:[14]

1949-54

1. Eliminate the most powerful people in the ancient regime.
2. Institute a land reform.
3. Experiment with the optimum size of agricultural production units.

1955-7

4. Gauge popular feeling through "letting the hundred flowers bloom."
5. Eliminate major detractors.

1958-60

6. Tap the industriousness of the people to test the limits of human capital and fervor.
7. Send foreign advisors home.

1961-5

8. Retrench, go back to private initiative and plots, reduce the size of production units, reestablish supply and marketing cooperative.

1966-9

9. Repeat step 6.

1970-4

10. Repeat step 8; become more open to the West and develop international trade.

1975-6

11. Restrict "bourgeois rights," strengthen the left.

1976-89

12. Go through a power struggle during the weakening and passing of Mao.
13. Purge left extremists.
14. Continue step 10.
15. Introduce the profit motive, specialization and trade, rural research, agricultural subsidies, responsibility system, and rural markets.

To take a place in world affairs commensurate with its population, the mainland giant that had responded lethargically to confrontation with the West in the 19th century had to awaken and move. A bouleversement of the ancient regime was necessary, and what better ideology to create the necessary uncertainty than the Hegelian dialectic. Despite the great discontinuities inflicted by swinging from stages of Type I to Type II, China's economy did swing upward. Mao realized that feelings of loyalty had to be transferred from the traditional family and village to the State. Even when transferred, these feelings could erode into complacency. Therefore, he indulged in periodic overkill to keep alive the new ethos. Any regime, to be successful, may have had to adopt a similar strategy.

Type I stages appealed to the traditional mysticism of Taoism and Buddhism, which held that all things were one, and that by meditating long enough (on the appropriate thought of Chairman Mao), one could achieve a state of harmony and resolve any conflict. There was also an appeal to relive the mass mobilizations that occurred during the traditional periods of the dynastic cycle. The object was to regain the mandate of heaven, not from a bad emperor, but from his "bad" advisors (such as Liu Shao-qi was accused of being), who had led the people into the latest misfortune. That it was important and culturally meaningful for Mao to call upon the age-old Chinese heritage is repeatedly manifest. In what other modern-day society could a Lin Biao be compared with Confucius (541-479 B.C.), Jiang-Qing with Empress Lu (206-195 B.C.), and Mao himself with Qin-Shi-Huang-di (221-205 B.C.)?[15] "The often spectacular contemporary changes and oscillations in China's society take place within a slowly changing millennial cultural mold."[16]

Type II stages appealed to another side of the Chinese cultural past: the industriousness, Confucian allegiance, and even the profit motive of the Chinese heritage. For example, at the time of liberalization in 1956-7, the regime instituted (1) free markets and flexible price controls, (2) private plots on up to 10 percent of the arable land, (3) a transfer of power to local government and individual enterprises, (4) a denial of Stalinist gigantism, (5) the reintroduction of the profit motive, (6) freedom of thought and slackening of the class struggle, (7) attention to quality not just quantity, and (8) an internationalist outlook. In 1961-5, there was also a subtle return to the ideas of Confucian allegiance to the throne: "Link one's family to Tian-an-men; cultivate according to local conditions but to fulfill state plans."[17] And the following agricultural goals could have been promulgated from Taiwan: high and stable yields, electrically-powered irrigation schemes, new stiff-stemmed high-yielding varieties, careful research, persuasion in extension, 7-hp tractors, consumer surveys, trademarks, advertising, birth control through late marriage, and spacing of children. The policy directives of the period after Mao's passing have been a revitalization of these strategies and an even closer rapprochement with the policies in effect in Taiwan.

The pendulum swing of mainland development from 1949 to 1989 harnessed the steady growth inherent in Confucian loyalties with the mystic fervor of egalitarian Buddhist-Taoism. In the name of forging a new China, the mainlanders actually drew heavily upon her past.

Policies Required for Taiwan

Taiwan in 1948-9 was a high-quality consolation prize for the Nationalists. The island was not untouched by war; the cities had been bombed, and there was work to be done in reestablishing communications and modernizing the cities. But the general state of the economy was better than on the mainland.

There were, however, two problems. First, local leaders had come to relish the independence they had enjoyed in the three years since the Japanese had left and planned to resist the take-over of government by the Kuomintang. Second, as on the mainland, there was inequality in the distribution of land. To redress these problems, the Kuomintang adopted a multistep strategy (numbered here to correspond to roughly similar stages on the mainland):

1. Eliminate the most powerful people in the ancient regime.
2. Institute a land reform.
3. Achieve optimum size of agricultural production and marketing unit through a system of cooperatives.
7. Send foreign advisors home.
10. Maintain private initiative and plots, optimal size of production units, openness to the West, and international trade.
13. Extend specialization and trade, rural research, and agricultural subsidies.

Of note are the fewer steps and lack of pendulum swings. The Taiwan regime decided (after eliminating its opponents) to follow the expert path and all but abandon the mystic approach to galvanizing the popular will. To be sure, the people's three principles (democracy, ethnicity, and well-being) were taught in the schools and used to name roads; there was a universal draft and national holidays; and the headlines reminded people daily of the eventual necessity of retaking the mainland. But these constant reminders were never transformed into political campaigns with overt confrontations. Buddhist-Taoist temples provided a mystical counterpoise on a private basis only.

The regime focused on steady progress and expertise, the free economy, private enterprise, and the shifting of population from the agricultural to the industrial sectors. Development was uneven because of this sectoral transformation, which reduced the percentage of the population in agriculture from over 60 to below 20 in the period 1949-89. There were four stages in the transformation.[18]

1. *1947-53: Recovery.* Land reform made more intensive use of labor and land, while chemical fertilizer and biological innovations were the major technological changes. In contrast to the method of expropriation on the mainland, land reform in Taiwan provided more economic incentives and proceeded in two phases: a 1949 rent reduction to 37.5 percent of the crop, which lowered real estate values and increased land sales; and a 1953 land-to-the-tiller program, which redistributed any land still above 3 ha of medium-grade irrigated and/or 6 ha unirrigated land.
2. *1953-68: Growth.* Mechanization began, largely through powertillers, encouraging migration to the industrial sector. This stage had three subdivisions, according to the goals of successive four-year plans:

   1953-6

   Self-sufficiency in the production of food, expansion of key industries (power, fertilizer, and textiles), and the development of new industries.

   1957-60

   More efficient use of natural resources and coordination of industry and agriculture (ironically, very similar to the avowed intent of the contemporaneous Great Leap Forward on the mainland, but with none of the excesses).

   1961-4 and 1965-8

   Development of export markets and increased investment incentives.
3. *1968-73: Slowdown.* So much labor had migrated to industry that there was not enough to maintain high agricultural growth. Cropping intensity declined, capital substituted for labor, and the major technological change was further mechanization. Within the general economy, the emphasis shifted from light to heavy industry.
4. *1974-89: Accelerated rural development.* With the shift to more capital-intensive means of agricultural production, farm production costs had risen and the income of farmers declined relative to urban income. A nine-faceted approach was announced to help maintain the growth rate in agricultural production. The rice-fertilizer barter system and the educational surtax on farm land were abolished, terms of agricultural credit were eased, markets and infrastructure were strengthened, rural industries were established, and the rural areas were divided up into specialized production and marketing areas to reduce the risk of and promote the benefits from specialization and trade. These policies were similar to those promulgated on

the mainland after 1979. In addition, such measures as joint corporations, cooperative farming, and a farm mechanization fund have been put in place in Taiwan to promote farm-size expansion and heightened factor efficiency.

Because of the preconditions of small size and facility of communications, Taiwan did not need so large nor so continued a push as the mainland. If the Communists had been in control, they might have performed in the same way, rebaptizing the above stages:

1. Restoring land to the masses.
2. The great leap forward (walking on two legs).
3. Building a proletarian worker-state (with due regard for the peasant anchor).
4. Supporting the peasant masses to strengthen the state.

Under the Kuomintang, this strategy worked because of the character of the people. Putting in long hours of their own will effectively doubled the labor force at constant food requirements. Willingness to cooperate in Farmers Associations and Specialized Production Areas—including the joint ownership of machinery—is a Chinese, not exclusively communist, trait all too evident in the highly successful cooperatives in Taiwan. Innovations in terms of multiple cropping, hog production, horticulture, and nutrient cycling all attest to the creativity and industriousness of a people *sans* Red Books. When savings were short, families and villages pooled capital, not unlike the "forced" pooling of capital in mainland communes and agricultural producer cooperatives.

The Organization of Rural Labor: Long-term Convergence

In Taiwan, the production, marketing, and irrigation of grain were formally organized through Farmers' Associations (FAs) and Irrigation Commissions (ICs). By the late 1960s, there were 364 FAs, mostly at the township (*xiang*) level, averaging 2200 households. Each township FA contained village-sized "small teams" of approximately 200-300 households. Even smaller units were composed of several households for "informal cooperation in harvesting... the result of friendship groups and family relations."[19]

The FAs provided extension information, credit, and livestock insurance; took charge of collecting, milling, and storing the grain each farmer was required to sell to the state; distributed nearly all the chemical fertilizer, seed, and farm implements; provided joint ownership and leasing of larger farm machinery; and furnished a marketing mechanism for producers. Membership fees were a nominal U.S. $0.14, and all farmers joined. The ICs were more fully autonomous and assessed water users' fees on an equitable basis.

Nongrain commodities such as pigs, poultry, and horticultural crops were informally organized and "raised by most farmers as a sideline but seldom as a principal endeavor"[20] through 1972. Then, when per capita incomes and urban population had risen, Specialized Production Areas were developed to provide these commodities regularly to city dwellers and at less risk and more profit to the producer.

On the mainland, rural labor organization followed the pendulum of the general economy but came to rest by the mid-1970s in a form that, despite different names, was quite similar to that of Taiwan. After land reform, Mao built up the level of decision making from "mutual-aid teams" of 7 to 10 households (corresponding closely to the informal cooperative-harvest groups in Taiwan) to the lower- and higher-level Agricultural Producers Cooperatives (APCs) of 50 and 246 households each (the latter corresponding to the "small teams" at the village level in Taiwan). By 1958, the average size of the APCs was 160 households,[21] much smaller than the FAs in Taiwan of the time. Thus, one major goal of the Great Leap Forward was to expand the decision-making unit to the township or *xiang* level (corresponding to the Taiwan FAs).

The expansion was carried out too quickly and some communes so formed were too large; but by 1974, there were 50,000 stable communes on the mainland. Each was made up of about 15 production brigades of 200 households (like the "small teams" at the village level in Taiwan), 100 production teams of 33 households each (corresponding to an enlarged version of the informal cooperative labor groups in Taiwan), and 3350 households (larger than average but no larger than many FAs in Taiwan). In 1980, these figures for the mainland remained stable.[22]

Not only was the structure of agricultural organization remarkably similar in both parts of China, the area farmed per commune/FA was 2,000 ha on the mainland and 2,200 ha in Taiwan! Tractorization had been a major feature of changing work patterns in both areas; between 1965 and 1973, the production of all tractors rose sixfold on the mainland, while the use of power tillers skyrocketed in Taiwan. Parallel leaps were made in both parts of China in the use of chemical fertilizer per ha.[23] The Communists, like the Nationalists, divided commodities into "first-category goods" (grain, oilseeds, cotton) produced on communally owned land, and "second-category goods" (poultry, pigs, eggs, and fruits and vegetables) produced on private plots, except for vegetable and hog production communes near the big cities. Finally, in both parts of China the agricultural wage fell below the industrial: in 1974 by 24 percent on the mainland and 33 percent in Taiwan.[24] Despite differences in ideology, the optimum system for organizing Chinese farmers seems strikingly similar.

Recent Hypotheses

Several authors have recently advanced hypotheses to explain the differing rural development performance of Taiwan and the mainland provinces with respect to institutional organization; prices, production costs, and farm income; and trade and competition. The following paragraphs not only define and clarify a subsample of these hypotheses, they also point to disagreements in interpretation that complicate the task of charting future Chinese rural development strategies.

Institutional Organization

Rural and General Economic Development. René Dumont, author of *Finis les lendemanins qui chantent*, a devastating critique of mainland China's failure to live up to the promise of combining egalitarianism with growth, has recently turned his pen to a work on Taiwan subtitled "The Price of Success." [25] After determining the keys to Taiwan's success that may be exported to other developing areas (reduction of inequalities between rich and poor and town and country, effective land reform, rural industrialization, reduced army and bureaucracy, controlled inflation, and an accent on education and small-scale industry), he hypothesizes that the environmental pollution, the fascism, and the open position of Taiwan on world markets have been the inevitable "costs" of this success. But Dumont's superficial acceptance of this hypothesis may be challenged, for many less developed countries have worse pollution, worse dictatorships, and/or more open positions on world markets, but lesser rates of economic growth and equity. Since these costs do not seem inevitably associated with Taiwan's successful policies, one might ask to what extent these latter may be modified and adapted to mainland China in the post-Mao period.

This is all the more so in that Taiwan's development has been remarkably equitable, with the ratio of the average income of the highest to the lowest quintile dropping from 15:1 in 1952 to only 4.6:1 in 1986. (The corresponding figure for rural mainland China in 1981 was estimated at 4.88).[26] Over the same period, Taiwan's literacy rate has risen to 92 percent, based on a minimum requirement of 9 years' formal schooling. The savings rate is between 35-40 percent, and it is estimated that Taiwan's economy will be the eighth most important in the world by the year 2006 (when Japan will be first).[27]

There seem, nonetheless, to exist other, real prices of Taiwan's economic success that the mainland provinces may wish to avoid. Kao argues that the destabilizing effects of industrialization, the using up of natural resources, the appallingly low readership of books, the high rates of crime and suicides, social tensions and fatigue, and the lack of a spiritual anchor in China's traditional philosophy, all lead to the danger of Taiwan's becoming an "economic giant, but a moral dwarf."[28] Unless, on a personal and national basis, Taiwan is willing to spend more in public goods, massive budgetary surpluses will not result in a better social infrastructure (transport, social insurance, basic research).

In light of this record, what projections can be made for mainland China, whose area is not only much larger, but which has espoused a socialist development philosophy? Mainland economists may be classified into three camps as to how they see the most likely development scenario for the year 2000.[29] Despite highly suspect data for the base period before 1977, the first school of thought maintains that policy has been the major source of rapid development, and that without a major new policy infusion similar to the production responsibility system, agriculture will flounder. A second, more optimistic view believes farmer self-determination has just begun to come into play, especially in rural nonagricultural production. A third and final view sees growth as saddle-shaped: the period 1978-84 restored the rural economy by releasing latent potential, but difficult post-1984 structural adjustments will necessarily reduce output growth rates, especially in grain crops. Nonetheless, by the end of this century, average living standards should reach **xiao kang** (U.S. $800/capita).

To help evaluate the respective merits of these projections, recent studies (e.g., Perkins and Yusuf, Wong) have attempted to explain the patterns of contemporary agricultural output growth on the mainland. Between 1950 and 1980, agricultural output enjoyed a highly favorable annualized agricultural growth rate of 3.2 percent, compared to the historical rate of 1 percent. These authors attempt to evaluate to what extent this growth was due to *direct* input increases in terms of land, labor, and such modern inputs as fertilizer vs. *indirect* organizational changes to use resources more efficiently. Mao Zedong seemed to believe that mass labor mobilization in rural China would tend to increase agricultural output without massive injections of modern inputs. As a result, billions of man-hours went into rural construction, deep plowing, backyard steel mill construction, etc. If Mao was correct, these measures have boosted output by increasing the productivity of land.

But Perkins and Yusuf present a second possibility: that labor mobilization may have had adverse impacts on agricultural output, and that it was mainly collectivization and other *organizational* measures that explain China's steady output growth. Assuming 16 percent labor productivity and zero land productivity increases, these authors calculate a maximum 14.7 percent increase in output due to organizational changes. Regardless of the relative merits of these two historical scenarios—which in fact merely seek to determine whether the Type I or Type II stages noted earlier had a greater development impact—Perkins and Yusuf tend to favor a strategy of more and better (modern) inputs to encourage future growth. Wong arrives at the same conclusion, but based on a different historical analysis. After comparing China with several other socialist economies for the period 1950-80, he calculates that China's superior agricultural performance was largely due, not to indirect organizational changes, but to direct increases in modern inputs. He finds that although the growth rate of land productivity outstripped that of labor, total factor productivity growth was negative. He concludes that China's agricultural performance was achieved at a very high cost of *extensification,* and had little to do with technical or institutional changes. Unfortunately, enough conjecture remains in the albeit creative analyses of both Wong and Perkins-Yusuf to cast considerable doubt upon the proper future path of China's rural development.

Macro Policies. Mainland China increased her GNP remarkably from 430 to 890 billion *yuan* in the period 1980-86, for a growth rate of 9.3 percent per year in constant *yuan*. In the same period, agriculture grew at 8.7 percent per year.[30] Unfortunately, beginning in 1984, inflation began to be a problem; for example, in the first 9 months of 1988, the price index rose by 16 percent over the same period in 1987. Webb hypothesizes several causes for this inflation, notably imbalanced growth rates among economic sectors, increasing consumer expenditures, budget deficit growth, easy monetary policies, increases in wage rates roughly double to real productivity gains in GNP, and artificial price hikes spurred by the double track pricing system.[31]

By contrast, occidental-style monetary and fiscal policies in Taiwan cut the inflation rate from 3400 percent in 1949[32] to 2 percent in the 1960s, to achieve an average rate of 8 percent for the period 1965-81.[33] Before such policies can be implemented effectively in the mainland, Tuan and Shi argue, the irrational price structure (notably heavy consumer subsidies to urban residents) and inefficient marketing and information systems must be changed.[34] Indeed, such seemingly effective short-term austerity programs as direct control over production systems, recentralized fertilizer distribution, a monopoly on cotton procurement, and severe cuts in loans and credit can not hope to promote sustained long-term agricultural development.

To make matters worse, the government decided to cut the share of overall capital construction that went to agriculture from 11.1 percent in 1979 to only 3 percent in 1988, at the same time as private investment became increasingly channeled toward consumer purchase.[35]

Prices, Costs, and Income

Agricultural Production. Mainland China is undergoing continued structural change in terms of the composition of crop area and the gross value of agricultural output. One projection forecasts that by the year 2000, only 54 percent of output value will come from crops, while livestock, forestry, sidelines, and fisheries will grow to 28 percent, 8 percent, 7 percent, and 3 percent, respectively.[36] However, without an effective marketing infrastructure, price mechanisms and macro policies (as exist in Taiwan), Niu, *et al.*, voice concern that the mainland may fail to effectively orchestrate this massive transition in output mix, while maintaining both the momentum of agricultural development and a healthy balance among physical resources, energy, and capital.

Similarly at the micro-level, Ma, *et al.*, have shown in a study in Jiangsu Province that, while farmers are generally technically efficient, they are not allocatively efficient because of the high risk associated with certain inputs and outputs.[37] These authors attribute this risk to continued price distortions, scattered holdings, weak market orientation, technical stagnation, and poor extension services. In addition, although farm household income has increased substantially since the inauguration of the production responsibility system, significant income differences remain, due to endowment in equalities in contracted farmland, machinery and draft animals; and unequal numbers of dependents per household. Finally, although specialized households tend to be better off, they do not seem to use society's resources more efficiently.[38]

Factor Markets. While markets for agricultural inputs in Taiwan are stable and well-developed, in part because they are coordinated by the Farmers Associations, much development is still necessary on the mainland. Not only is supply unreliable, but cost-push inflation increased their cost by 17.9 percent in the period 1978-84.[39]

One major area of weakness is in capital markets, intensified by reduced government investment in agriculture and the lack of new banking institutions to replace capital allocation mechanisms that existed under the communes. Feder, *et al.*, conclude that formal and informal credit markets on the mainland are highly segmented and do not substitute effectively for each other from the point of view of the potential borrower.[40] This is because informal credit is often used for nonproduction purposes and nonfungible projects. In addition, these authors report a link between output market development and increasing demand for production credit. But productive investment remains substantially lower than investment in housing because of small farm size, insecure land tenure, and inadequate financial arrangements. All these constraints can be reduced by appropriate policies (notably new legal mechanisms in land markets).

Similar problems plague labor markets on the mainland. Imbalanced sectoral investment and a lack of infrastructure in rural towns and cities have reduced society's ability to absorb surplus village labor.[41] Slogans to "leave the land and the countryside" and "use small towns as a development outlet" have not been implemented in a stable enough manner to fully use rural labor. Rather, vertical and horizontal links similar to those that have evolved in Taiwan seem necessary to unify villages, small towns, and medium-sized cities.

Trade and Competition

Comparative Advantage. During the Mao years on the mainland (1949-76), each province was encouraged to ignore comparative advantage and become self-sufficient in agricultural and industrial products to better resist possible renewed foreign aggression. But since 1979, Chinese leaders have promoted limited specialization and favored the principle of comparative advantage in the selection of crop and livestock commodities for local production, rural processing, and interregional and international trade. One current issue is to what extent China since 1979 should be searching for new patterns of regional specialization and trade along the lines that Taiwan has been able to follow uninterruptedly since before 1949, and to what extent it can return to the old pre-1949 mainland patterns. What makes the question more challenging is that relative prices and factor productivities have changed, so that the original patterns may no longer be an appropriate guide to new-found patterns of specialization.

Two studies—one from before and one from after the Mao period—provide at least a partial response to this question. The first is Buck's 1929-33 survey of 16,786 farms in 168 localities and 22 provinces.[42] The second is Webb's ranking of 1987 planted area, profitability, and production cost by crop and by region.[43] Webb divides China into 7 regions, using a nomenclature similar to Buck's, who nevertheless ignored the pastoral and extreme northern areas of the mainland and settled on 8 (rather than 7) major crop producing regions, of which 3 are wheat-based in the north, and 5 are rice-based in the south.

A comparison of the multiple cropping index, the percentage area planted, and the degree of specialization (combined percentage area of the two principal crops in each region) between the two studies suggests the following conclusions. First, the broad agricultural policy and rural construction trends since 1949 on the mainland have increased the MCI significantly, except perhaps in the northeast and the northwest, where the potential to do so was limited. [44]

Second, within this rising cropping intensity, there have been significant shifts in the relative area cultivated to specific crops. Notably rice, wheat, and millet have declined significantly everywhere, while corn and soybeans have risen markedly in the northeast. Cotton has declined in the northwest and the east, while remaining about constant in the north. The net result of these shifts has been an increasing degree of regional specialization 1933-1987, except perhaps in the northwest.[45]

What is interesting is that, according to Webb, the current patterns of specialization in crop area correspond also to the present ranking of individual crops in terms of both net returns and production costs per *mu*,[46] demonstrating that China is not simply returning to the principle of comparative advantage, but also finding up-to-date patterns of regional specialization for trade.

International Trade. Exports represent 55 percent of Taiwan's GNP (up from 17 percent in the period 1961-65), of which 48 percent are sold to the United States. On the import side, 33 percent of imports come from Japan. Taiwan is thus highly susceptible to U.S. business cycles, especially since 80 percent of American exports come from 250 companies.[47]

For the much larger and more climatically diverse mainland, there are no immediate answers as to how internal agricultural markets can be more efficiently linked to international markets or even nonagricultural domestic markets. Because of inadequate capital investment in the transportation infrastructure, Niu, *et al.*, argue that China's international comparative advantages have not yet been realized. In noted contrast to Taiwan, the mainland's huge population and limited industrial development severely restrict import payment capacity; and a strategy of coarse-grain imports to develop the mainland's livestock sector does not appear wise. It also seems obvious, again unlike Taiwan, that the mainland can neither continue to import large quantities of grain in the long term nor become a significant grain exporting country. Rather, exports and imports will be used to achieve partial adjustments.[48]

For this reason, recent projections (e.g., Ralston, Carter, and Zhong) cause concern over the growing dependence of mainland China on imports, notably of wheat. For example, Ralston in 1983 forecast three scenarios (10, 17, and 22 million metric tons) for 1990 wheat imports based on differing assumptions as to population growth, consumption levels, and agricultural productivity. Already for the year 1989, the level of wheat imports has exceeded 15 million tons.[49] However, 1989 was a record year in terms of grain output (407.5 million tons), and concern remains that

1991 and beyond may return to the stagnation of the years 1984-88. As a result, the mainland's ministry of foreign economic relations and trade predicts a 100 percent increase in agricultural import costs between 1989 ($6.7 billion) and the year 2000 ($12-13 billion), with wheat imports rising to close to 20 million metric tons.[50]

Conclusions

Despite different paths taken, the current rural structures of Taiwan and mainland China bear striking similarities. What differs is the macro-economic, market and international trade environment in which agricultural production and rural transformation must continue to evolve. While the proximate cause of these environments is historical ideological differences, such differences may themselves be largely explained by the geographical determinism of the ethos necessary to galvanize each of China's two parts. The fact that these ideological differences are now diminishing indicates a heightened possibility of convergence between Taiwan and the mainland in terms of market structure and development polices.

The remaining chapters of this book will present factual information and further interpretations of these and other hypotheses. Taken together, they will lend deeper insight into the likelihood of continued rapprochement in patterns of Chinese rural development with respect to institutional environment; prices, markets, and income; and international trade and competition.

Notes

1. An earlier version of the first part of this chapter appeared as an article by the same author in *Asian Profile*, Vol. 9, No. 3, June 1981, titled "The Myth of Two Chinas: An Agricultural Perspective."

2. Taiwan Provincial Department of Agriculture and Forestry, *Major Indicators of Taiwan's Agriculture* (Taipei: Provincial Government Press, 1976), p. 1.

3. Jan S. Prybyla, *The Chinese Economy - Problems and Policies* (Columbia: University of South Carolina Press, 1978), p. 31.

4. Despite some spurious gains during the Great Leap Forward, 1973 irrigated area was five times that of the early 1950s. Jan S. Prybyla, p. 32.

5. Ching-yuan Lin, *Industrialization in Taiwan* (New York: Praeger, 1973), p. 37.

6. Data from Taiwan Provincial Department of Agriculture and Forestry, p. 1; and Jan S. Prybyla, p. 30.

7. Neil H. Jacoby, *U.S. Aid to Taiwan* (New York: Praeger, 1966), p. 51.

8. Choh-ming Li, "China's Industrial Development, 1958-63," Choh-ming Li (ed.), *Industrial Development in Communist China* (New York: Praeger, 1964), p. 30-1.

9. Ching-yuan Lin, p. 13.

10. George Barclay, *Colonial Development and Population in Taiwan* (Princeton: Princeton University Press, 1964), p. 36.

11. Jan S. Prybyla, p. 30.

12. Jan S. Prybyla, *The Political Economy of Communist China* (Scranton, Pennsylvania: International Textbook Company, 1970), p. 29.

13. However, the actual difference in size of land holdings between the small and large operators was not nearly as great as the Communists would have one believe (Jan S. Prybyla, 1970, p. 35).

14. The following sources contributed to this listing of stages: Jan S. Prybyla, 1970 and 1978. Zhan Wu, "Wan-jeng zhi-xing, nong-lin-mu-fu-yu bing jufang-jen jia-kuai nong-ye xian-dai-hua bu-fa" ("Completely Implementing the Tool of Jointly Supporting Agriculture, Forestry, Herding, Agricultural Industries, and Fisheries' as a Means of Hastening the Modernization of Agriculture") Jing-ji Yen-jiu (Economic Research) Vol. 2 (February, 1979) pp. 30-36. Benedict Stavis, "Turning Point in China's Agricultural Policy," Michigan State University Rural Development Working Paper (East Lansing: Department of Agricultural Economics, 1979).

15. Jan S. Prybyla, 1978, p. 4.

16. Ibid.

17. Jan S. Prybyla, 1970, p. 359.

18. The following sources contributed to this listing of stages: Erik Thorbecke and Jackson Karunasekera, "Technological Change and Growth monograph (Ithaca, 1978). Terry Y.H. Yu, "The Accelerated Rural Development Program in Taiwan," *Industry of Free China* (March, 1977).

19. Frederic H. Chaffee, *et al.*, *Area Handbook for the Republic of China* (Washington: U.S. Government Printing Office, 1969), p. 268.

20. Chaffee, *et al.*, p. 258.

21. Jan S. Prybyla, 1970, p. 283.

22. *China Encyclopedia*, as cited in Perkins and Yusuf, p. 93.

23. Jan S. Prybyla, 1978, p. 185; Taiwan Provincial Government, *Taiwan Agricultural Yearbook for 1977* (Taipei: Provincial Press, 1978).

24. Jan S. Prybyla, 1978, p. 56; Terry Y.H. Yu, p. 15.

25. René Dumont, 1987, *Taiwan: Le prix de la reussite*. Editions de la Decouverte. Ville Saint-Laurent (Quebec).

26. D. Perkins and S. Yusuf, 1984, *Rural Development in China*. Johns Hopkins University Press. Baltimore, p. 144.

27. These statistics come from Charles H. C. Kao. Undated mimeo. "Taiwan Development Experience: Lessons and Perspectives," 14 pp.

28. Ibid., p. 6.

29. The portrayal of these schools draws on Niu Ruofeng and Peter Calkins, p. 449.

30. Webb Shwu-eng H., 1988, "Inflation in China Since 1978 Economic Reforms." *CPE Agriculture Report*, Vol. 1, ERS/USDA, Washington, DC, pp. 37.

31. Ibid., pp. 37-8.

32. Dumont, op. cit., p. 23.

33. Tibor Scitovsky, 1986, "Economic Development in Taiwan and South Korea, 1965-81. In Lawrence Lau and Lawrence Klein, eds., *Models of Development:*

A Comparative Study of Economic Growth in South Korea and Taiwan. ICS Press, San Francisco.

34. Francis C. Tuan and R. Shi, 1990, "PRC Agricultural Trade Development and Prospects: With Emphasis on Pacific Rim Countries." Paper presented at the International Agricultural Trade Research Consortium, Hawaii, August 1-2, p. 4.

35. Ibid., p. 15.

36. Niu and Calkins, op. cit., p. 446.

37. Nathaniel Ma, Peter Calkins, and Stanley R. Johnson, 1989, "Technical and Allocative Efficiency vs. Equity in Shuyang County, Jiangsu: General Households vs. Specialized Households in John Longworth, ed., *China's Rural Development Miracle*. University of Queensland Press, St. Lucia.

38. Ibid., p. 363.

39. Niu and Calkins, op. cit., p. 448.

40. Gershon, Feder, Lawrence Lau, Justin Lin, and Luo Xiaopeng, 1989, Agricultural Credit and Farm Performance in China." *Journal of Comparative Economics*.

41. Niu and Calkins, op. cit., p. 448.

42. John L. Buck, 1937, *Land Utilization in China*. University of Nanking, Nanking.

43. Shwu-Eng H. Webb, 1990, "Regional Agricultural Development in China, 1978-88: Parts I and II," *CPE Agriculture Report*, Vol. II, Nos. 2 and 4. ERS/USDA, Washington, DC.

44,45. Because Buck and Webb's regional definitions differ the most for these regions, it is difficult to say.

46. The two exceptions are rice in the Northwest and North and cotton in general, for which high production costs often discourage farmers from expanding area planted and exploiting high returns per mu on local markets.

47. These statistics are taken from Kao, op. cit., pp. 9-10.

48. Niu and Calkins, op. cit., p. 447.

49. Tuan and Shi, op. cit., Figure 4, p. 5a.

50. Ibid., p. 19.

References

Carter, C.A. and F. Zhong. 1988. *China's Grain Production and Trade: An Economy Analysis*. Westview Press, Boulder, Colorado.

Niu, Ruofeng and Peter H. Calkins. 1986. "Towards an Agricultural Economy for China in a New Age: Process, Problems, Response, and Prospects. *American Journal of Agricultural Economics*, Vol. 68, No. 2, May.

Perkins, D. and S. Yusuf. 1984. *Rural Development in China*. Johns Hopkins University Press. Baltimore, Maryland.

Ralston, K.L. 1983. *Potential Chinese Wheat Imports 1985 and 1990*. M.S. Thesis, Cornell University.

Webb, Shwu-Eng H. 1988. "Inflation in China Since 1978 Economic Reforms." *CPE, Agriculture Report*, Vol. I, pp. 37-39. ERS/USDA, Washington, DC.

Wong, Lung-fai. 1986. *Agricultural Productivity in the Socialist Countries*. Westview Press, Boulder, Colorado.

3

Economic vs. Noneconomic Factors in Chinese Rural Development

D. Gale Johnson

The tragic events of mid-1989 in the People's Republic of China make it clear that we should never ignore noneconomic factors in considering any aspect of life. One cannot escape the fact that a group of octogenarians was willing to impose enormous costs upon the people they represented to maintain their power and to prevent the sharing of that power. The students who led the democratic movement can be criticized for the lack of concreteness in their objectives, but it is equally difficult to determine what (except for memories and pride) caused these old men to inflict such pain and hardship upon the citizens of China.

I will not dwell upon these tragic events except to note that over the past two years I have felt that the rural reform's enormous achievement of giving greater freedom and economic independence to farm people could be greatly eroded and even largely lost. If what many of us thought never could happen in Tienanmen Square did happen, it is also possible that the local leaders could regain the political and economic power they lost when the responsibility systems were introduced and the communes were abolished. The village, township, and county leaders will be successful in regaining their power if they have the support of provincial and national leaders, and I shall demonstrate that such support cannot be ruled out.

This chapter will consist of three main parts. The first will be my evaluation of the consequences of the 1979 and subsequent reforms. The second will emphasize what further steps need to be taken to continue the progress already achieved. The final part will emphasize my concern that the gains in freedom and independence of rural people will be compromised under the guise of solving some of the important problems that confront agriculture today.

A Positive Evaluation of the Reforms

A number of issues arise in evaluating the rural reforms undertaken since 1979. One set relates to the combined consequences of all the reforms with respect to important objectives such as aggregate output, the mix of output, resource productivity, and farm family income. Another set deals with the effort to identify the relative effects of particular components of these reforms, especially the household responsibility system and the removal of restraints on the use of the resources of farm people in nonagricultural activities.

In the last two years, a growing chorus has claimed that the mainland's agricultural reforms were no longer succeeding. Its soloists have included numerous prominent Chinese officials and scholars as well as some Western observers. The criticisms appear to rest primarily upon three statements of fact. First, while the rate of growth of agricultural output was high from 1979 through 1984, it has declined significantly thereafter. Second, grain output since 1984 has been less than the 1984 level. Third, the retail prices of agricultural products for which price controls were abolished have increased very sharply. These statements, while factually correct, say nothing about the success of the rural reforms.

The gross agricultural output for 1979 to 1984 grew at the remarkable compound annual rate of 7.7 percent. This figure is based upon the most recently revised official data that no longer counts village industry as a part of agriculture. For the period 1985 to 1988, the growth rate did decline, but to a very respectable 4.0 percent annually, significantly above the population growth rate of about 1.5 percent.

The output of grain increased rapidly from 1979 to 1984, reaching a peak level of 407 million metric tons. The next closest year was 1988, with a production of 404 million tons. The grain yield per unit of land in both 1987 and 1988 exceeded that of 1984; the area devoted to grain declined slightly. Some observers, especially Chinese, want to use grain output as the criterion for judging the annual performance of agriculture. Outside of jitteriness engendered by the years of famine under Communist rule, there is no reason why a *yuan* of grain output has any more value than a *yuan* of pork or cabbage output. Because Chinese leaders have kept the grain price well behind prices of other important farm products, it is disingenuous of them to use the failure of grain output growth as a sign that the rural reforms themselves have failed. While the prices of several important farm products, such as pork, poultry, and fresh vegetables, are now largely determined by the market, a large percentage of the grain is still marketed at a price determined by the government. Since the retail price of rationed grain is fixed, any increase in grain procurement requires a larger governmental subsidy.

There are difficulties in interpreting data on prices received by farms. But if mixed average purchasing prices are accepted as an approximation of prices received by farmers, grain prices have lagged behind those of livestock and poultry since 1978. The nominal grain price increased by 93 percent between 1978 and 1987, while hog and egg prices increased 155 and 120 percent, respectively (State Statistical Bureau, 1989, p. 708). If grain has such high priority, the government should raise its real price sufficiently to elicit a higher level of output. Or, recognizing that it may be cheaper to obtain marginal quantities of grain from the international market, the government should stop evaluating the reforms on this single and misleading criterion.

Raising real grain prices is only optimal in the short run; the longer run solution is much more complex. If China is to produce increasing quantities of grain at constant or, preferably, declining real prices, ideological or noneconomic factors must come into play. At a minimum, there must be an efficient means of enlarging the size of crop operations and of assuring the security and transferability of the right to use (if not to own) land. But at present, after a decade of rural reform, numerous issues touching the ownership and use-transfer of land remain unsettled. I shall elaborate on this point later.

There is disagreement about the factors responsible for the remarkable increase in agricultural output and productivity since 1978. The reforms were not organized to help the economist attribute the gains to particular elements of that reform, for many changes occurred simultaneously or in rapid succession. The price at which the government acquired farm products increased by 30 percent between 1978 and 1980, in part because higher premiums were paid for deliveries in excess of the required quota. The major institutional changes that introduced numerous responsibility systems and finally abolished the communes were started in 1979 and essentially completed by the end of 1983. While there were many different forms of the responsibility systems at first, rather soon the household responsibility system emerged as the nearly universal form. But other major policy changes occurred as well. These included encouraging rural trade fairs, permitting farmers to sell directly to urban residents, enlarging the scope of accepted sideline activities to include almost any nonagricultural activity, and allowing individuals to buy and sell products with the object of making a profit ("unearned income," as defined by Marxists). These nonprice, nonresponsibility-system changes made possible more efficient use of labor in agriculture by permitting hitherto prohibited employment outside of agriculture and great improvements in the scope and efficiency of the marketing system.

It is difficult to systematically disentangle the effects of these individual policy components upon increased agricultural output and productivity.

Although the major contenders for credit for the increase in agricultural output are prices and the responsibility system, few authors have conducted detailed studies. I know of three: Justin Lin; John McMillan, *et al.*; and James Wen. Lin's study, the earliest, was based upon a production function approach using pooled time series and cross section provincial data. To represent institutional change, Lin used the percentage of the households in each province that had adopted the responsibility system for each year from 1980 to 1983. He estimated that 45 percent of the increase in agricultural output between 1980 and 1983 was due to the adoption of the household responsibility system (Lin, 1986). Notably, this leaves a significant percentage of the total increase in output from 1979 to 1988 to be attributed to other factors, such as price increases and improved technology. Of the total increase in agricultural output of 82 percent from 1979 to 1988, Lin's analysis would attribute less than a fifth of the increase to the responsibility system—17 percent of it to be exact.[1] I believe that Lin's results underestimate the effects of the responsibility system upon output since he assumed that all of the effects would be realized in the year of adoption. I think it is reasonable to assume that the farm households did not realize the full benefits of their managerial freedom in the first year but that there were some effects of learning by doing. However, I should add that Lin's county-level data supported his hypothesis that all of the benefits were found in the year following nearly complete adoption.

John McMillan, John Whalley, and Li Jing Zu estimated the effect of the household responsibility system on the increase in total factor productivity from 1979 to 1984 by decomposing the increase into a price component and an incentive component. Their analysis assumed that the primary effect of the household responsibility system was to heighten the incentive to work. By their estimates, total agricultural output increased by 62 percent from 1979 to 1984, while total factor productivity increased by 41 percent. They estimated that the responsibility system could be credited with 78 percent of the growth in total factor productivity and 52 percent of the growth in output. In their model, price increases not only accounted for 22 percent of the increase in total factor productivity, but were also a factor in the substantial increases in both current and capital inputs.[2]

James Wen used a different approach, namely the estimation of a supply function for total agricultural output. Rather than using the retail price of industrial products sold to agriculture—a series that is clearly unrepresentative of the production and living costs that farmers face—Wen used a price index derived from the measures of consumption expressed in current and constant prices. The coefficient for the household responsibility system indicated that it permanently increased farm output by 56 percent. On a comparable basis, Lin's estimate gives a permanent output increase of 14 to 20 percent while McMillan's yields 32 percent.

While these three different analytical and empirical approaches give somewhat different estimates of the significance of the household responsibility system in explaining output and productivity increases, each indicates that there has been a significant effect. Cursory examination of annual changes in factor productivity growth supports the hypothesis that the household responsibility system had a significant impact. Using a measure of total factor productivity based on logarithmic weights, annual rates of growth increased from 4.8 percent in 1981 (the first year after the significant increase in the responsibility system) to more than 8 percent in both 1982 and 1983 and 14 percent for 1984.[3] As the latter year enjoyed relatively good weather, this large increase needs to be discounted by an unknown degree. The average rate of productivity growth for 1985 to 1988 declined to a very respectable 2.6 percent, somewhat more than the U.S. rate for the decade ending in 1987.

Further Desirable Measures

Successful as the rural reforms have been, some problems could have been prevented had more thought been given to the governmental and economic consequences of abolishing the communes and had the government kept its commitment to invest in agriculture and the rural areas. Instead of increasing from 11 percent of total state investment to 18 percent, as called for by the Central Committee in 1978, such investment declined to only 3 percent of the total by 1987.

The communes had important governmental as well as economic functions. When the communes were abolished there did not appear to be a blueprint for the transfer of governmental functions to specific governmental units. As a result, certain functions, such as the maintenance of irrigation facilities, were largely left undone. The communes had supported several useful institutions, such as schools, medical clinics, and hospitals, by using part of communal income. When the communes were abolished, these services, which the farm people had thought to be free, now had to be paid for by taxes. Consequently, in some unknown fraction of local areas, schools and medical services suffered.

The abolition of virtually all prohibitions on nonagricultural economic activities in rural areas was a pivotal step with major consequences. Nonagricultural employment in rural areas has increased from about 30 million before the reforms to more than 80 million now, an increase equal to half of U.S. employment in 1990. Almost all of these jobs have been generated by local ingenuity and resources, for national and provincial governments have invested nothing in such enterprises. As of 1988, the gross value of output of rural industries has exceeded the gross value of farm production. A significant part of the increase in rural family incomes has been due to this large transfer of labor out of farming into a more productive activity.

But, probably, the most important and difficult challenge is to increase the size of farm operations. There are now about 200 million households engaged in farming; on the average, each has somewhat more than an acre of cropland. As economic modernization occurs, the number of farm households will decline. But currently there is no national policy to facilitate the transfer of land from one household to another. Nor is there an accepted legal designation of the ownership of the land. It is not clear what rights the production team members have in the land they have farmed for many years nor what authority the village has over the land. Or does the land, certainly *de facto* if not *de jure*, belong to the state?

Rights to use the land have been assigned to households. But it remains unclear whether those rights are attached to specific parcels of land or only to a certain share of the land in the production team area of the village. In the latter case, the farmer has little incentive to make long-term improvements in the land since, after a period of 3 years, land is reassigned to take account of changes in the composition of families and of the population in the village. A few communities have decided not to reassign land as population changes, but they remain a small minority. Because the economic benefits are great of ensuring clear title to whatever rights have been assigned, the failure to take the necessary measures is very likely due to nonmarket or ideological factors. Land is clearly a means of production; can it or should it be privately owned? If other means of production, such as tractor and trucks, can be so owned, why not land? Is it appropriate to sanction families giving up farming and renting their use rights into the indefinite future? These are questions for which there have been no answers in a decade.

I have estimated that the real incomes of farm families approximately doubled between 1978 and 1986 (Johnson, 1987, p. 29). Nonetheless, the urban bias that for so long supported urban incomes at 2.5 or so times rural incomes has persisted. Urban consumption expenditures measured in *yuan* have been about 2.5 times rural expenditures since 1952, with an upward trend to 2.9 times in 1977 and 1978, a declining trend to 2.2 times in 1983 and 1984, and a final increase to 2.5 in 1987. But these data do not fully reflect the differences in real consumption. While almost no consumption items are subsidized in rural areas, housing, household energy, urban transportation, and other products are heavily subsidized in urban areas. The contrast in government largesse is readily evident in housing. In 1987 urban residents allocated 0.9 percent of their living expenditures for house rent and 1.2 percent for water and electricity for a total of 2.1 percent (State Statistical Bureau, 1989, p. 717). Rural people, for whom there are no housing subsidies, devoted 14.5 percent of their consumption expenditure to housing plus 4.8 percent for fuel in 1987 for a total of 19.3 percent (State Statistical Bureau, 1989, p. 733). True, they had about twice as much housing space as their urban counterparts, but they paid dearly for it.

There are two other policy biases that have not been ameliorated and may have, in fact, become more striking over the past three decades. One is the anti-rural investment bias in favor of urban areas and industry. Even before the 1979 reforms, only 11 percent of centrally-directed investments went to agriculture and the rural areas; this was for 80 percent of the population. As noted, the central government essentially abandoned any pretense of investing in rural areas by the late 1980s. Much of the governmental investment in recent years has gone to make life more comfortable for the urban bureaucrat in the form of new apartments and new office buildings. In spite of this neglect of rural investment, the Communist Party now expects the rural areas to support its current oppressive campaign against those who wanted some modest changes in the political system.

The third bias is coastal. Mao did try to emphasize investment in the interior areas, although his efforts had little effect in improving the lot of the people and were probably motivated by his paranoia of military invasion. In spite of both the concentration of investment in the coastal areas and the numerous locational and infrastructural advantages of such areas, there does not seem to have been a sharp widening of the income differences in rural areas. It is quite remarkable that the poorer provinces, where there may have been no increase in real income for two decades and famine had not been conquered, were able to achieve rapid growth in real income after 1978 without a significant increase in resources except for that most precious of all: independence and the freedom to make many of one's own decisions.

I noted that there has not been a sharp widening of rural incomes between the low-income interior provinces in the northwest and southwest compared to the coastal area. However, there was some widening for the period from 1978 to 1985 for the five provinces out of ten for which 1978 data are given (State Statistical Bureau, 1989, p. 736). Only one province had an increase in income greater than the national average increase in rural incomes. For 1980 to 1985 the record is much better. Assuming that a difference in income growth of 5 percent from the national average is within the margin of error, six of the ten provinces had the same percentage increase, one had a greater increase and three had lower ones. By this rough measure, there was some deterioration in the relative position of the western provinces between 1985 and 1987 as the importance of nonagricultural incomes of farm people increased, especially in the coastal areas. For 1980 to 1988, none of the western provinces had an income increase greater than the national average, four had the same percentage increase and six had a lower rate.

A form of blatant discrimination against rural people is found in the severe limitations on permanent migration from rural areas to the larger cities. But because of the demand for construction and generally unskilled workers in the large cities, millions of individuals from rural areas have been permitted to go to the cities as contract workers. A contract worker has no job security, receives minimal benefits in terms of health insurance, and does not receive ration coupons to buy subsidized foods. Nor do such workers receive the per capita subsidies that have been given to regular urban workers to compensate for higher prices for vegetables and meats. Thus, at most, contract workers receive the same money wages as urban workers and none of the many benefits. And when there are restraints on such economic activities as construction in 1989, and 6 million construction workers will lose their jobs, most of these will be from rural areas. Yuan Mu, as spokesman for the State Council, after noting that most of the construction workers laid off were from rural areas and that the few from the cities would be given either new jobs or retrained, concluded: "Therefore, China's overall unemployment rate will not increase" (Yuan). Apparently, nearly 6 million rural workers who will lose their jobs don't count since they will leave the cities and vanish into the countryside!

Will Rural Freedoms Be Restricted?

Another way of putting the question is: Will the household responsibility system be emasculated? I raise this question, not in the context of the horrible events of 1989, but in response to certain concrete deviations that have already occurred from the household responsibility system. Whether or not it is national policy, certain political forces are moving to recentralize the power that once rested in the hands of production team and brigade leaders and the officials of the communes. In most parts of rural China, it appears that the individuals who under the commune system had a great deal of authority still have it in the leadership of villages and townships. These remain the groups that make the key decisions affecting farm people: how delivery quotas are allocated among families, farm inputs are made available, and incomes from village and township enterprises are distributed.

The stated claim that the responsibility system has not solved the grain problem becomes a rationalization for such individuals and groups to seek alternative means of increasing grain production. The failure to provide the optimum circumstances for enlargement of crop operations under the responsibility system becomes an excuse for partial recollectivization of agriculture.

I have argued elsewhere (Johnson, 1987) that, as the value of farm labor increases with economic growth, China will have to choose among three major alternatives. One alternative is the path chosen by Japan and later South Korea; namely, to increase real farm prices enough to make it possible for farm families to earn a reasonable level of income on farms only slightly larger than current ones. A second is to permit the transfer of the rights to use land—either by rental or sale—so that the land area of operating units can be enlarged over time within the framework of the household responsibility system. The third is to sanction and/or encourage collective forms of farming, with those activities where there are economies of scale being done collectively, while other activities requiring attention to detail and careful work are left to each separate family.

The first alternative (to have farm prices significantly above world market prices) is not viable for China. When such policies were adopted in Japan and South Korea, farm people represented a minority of the total population, and the rest of the population had sufficient income to pay domestic prices that exceeded import prices. In China, the nonfarm minority of the population cannot conceivably pay prices high enough to make the present-sized farm units economically viable for the next decade or longer if rates of national economic growth are at 5 percent or more.

I give three examples that may reflect what is in store for agricultural organization in the years ahead. I do not know how typical these examples are or how likely they are to be duplicated in the future. If these different forms of agricultural organization were arrived at *democratically*—meaning that within a village people could freely choose how their agricultural resources were to be organized—there is nothing inherently inappropriate about any particular form of organization. But given the power and authority held by village and township leaders in most parts of rural China, one has a right to be skeptical about the degree of freedom of farm families to choose how they wish to farm. These choices include staying with the responsibility system, accepting a return to hired worker status, or being subject to partial or full collectivization.

The examples I give appear to have been supported in Document No. 1, issued by the Central Committee of the Communist Party on January 1, 1986. While stating that "it was imperative to adhere to integrating unified methods and decentralized ones," it provided the opening for centralized decisions:

> As there are considerable disparities in social and economic conditions in different places, the content, forms, scale and degree of combined unified with decentralized management may vary. In regions where the collec-

> tives have less accumulation, production is concentrated on similar crops and intended mainly for local consumption; the work should start at the root to earnestly help the peasants solve their difficulties in production and circulation and steadily enrich the cooperative content. In places which have a well-developed economy and soundly based collective enterprises, the advantages of unified management and distribution should be used to the full, agricultural capital construction and technical transformation strengthened, operational scale appropriately adjusted and coordinated growth of agriculture, industry and commerce encouraged (Central Committee, p. 7).

While the document states that the responsibility system should not be changed "at random, departing from the people's wishes," it added that "people are dissatisfied with localities that have not undertaken tasks that one household cannot do well or at all." Within these guidelines, local officials seem to have considerable room for modifying the household responsibility system.

In a report on the National Rural Work Conference held in November 1988, Zhao discussed the conclusions that had been reached with respect to the improvement of the household responsibility system. One conclusion was that there should be the development of socialized services "which are needed before and after production, and cannot be provided, or properly provided by a single household, such as information, technological assistance, funds, crop protection services, machines, warehouses, and circulation services" (Zhao, p. 41). These proposals seem consistent with individual responsibility.

The above was followed by:

> As for large-scale management of land, several basic points were clarified at the meeting. 1) It is conducive to raising land and labor productivity, and solving the problem of lower returns from growing grain. 2) There are two prerequisites for large-scale management: job opportunities are available for surplus labor forces, and there is a certain amount of accumulation funds. These two conditions are not found in most areas of the country at present. 3) In places where conditions are ripe, we can encourage and advocate large-scale management. But first we should proceed at an appropriate pace and engage in intensive farming. Second, we should stick to the principle of voluntary participation, not resort to coercion and commandism, and not act with undue haste. Third, we should not slip back to the old rule of allowing everyone to eat from the same big pot.

Local officials, on reading these conclusions, can readily justify substantial increases in township or village authority over the organization of farming.

It is presumably their decision whether there has been adherence to voluntary participation and everything has been done in "due" haste.

The first form of organization I shall describe is that of the Big Family.[4] This type of household has been assigned land and perhaps other resources by a village or township and employs significant numbers of hired workers. In a village near Shenyang that I visited in 1985 there were 650 farm workers. Ten of the families operated small family units, using only their own labor, under the household responsibility system. The remaining 625 workers, an average of 37 workers per farm, received a wage plus a share of the net income. In this village, where vegetables were produced under plastic, I met with one of the 17 Big Families who operated 7.5 acres of land, of which 5 acres were under plastic, and employed 52 farm workers. Another Big Family hired 80. The family that I visited also controlled a factory employing 56 workers. In the combined farm and factory, this family employed 108 workers or 7 percent of all the workers in the village. The member of the Big Family that I met with had been the head of a production team in the immediate past. This village had given up collective agriculture only the year before.

How prevalent is the Big Family? I was told in Shenyang that 5 percent of all farmland in China was operated by Big Families, but I have no confirmation of this number. I have asked others about it; those who had a view said that the 5 percent seemed reasonable as of 1985.

A second example of significant collectivization of the agricultural operations occurs in about 5 percent of the villages of Zouping County in Shandong Province as of 1987. The approach to farming was called the Five Unifieds. What this meant was that while land had been assigned to each household, five farming operations were carried out collectively. These included plowing the land, seeding, fertilizing, irrigating and harvesting the wheat. The other crops, principally corn, cotton, and peanuts, were harvested by the households since these latter crops involved hand labor. Village officials operated and controlled nonagricultural enterprises, including manufacturing and processing plants. The profits from these enterprises were used to subsidize certain farming operations, such as plowing, irrigation, and fertilization, and to provide free fish and ducks for certain festivals or holidays. In this way, significant incentives were provided to induce the villagers to accept the incursion of collective controls in the agricultural sector. This approach to farming also meant that the crops grown by each family were decided not by them but by the village authorities. I do not know how satisfied the farm families are with these arrangements. Clearly, village officials have a great deal of authority due to their control of the nonagricultural activities and the direction of agricultural operations. It must be recognized that the Five Unifieds include operations in which there are either economies of scale, as in plowing the fields with a tractor, or need for unity of decisions, as in operating the irrigation system. So far as I know, there have been no published studies of the economic effectiveness of this system compared to that of the household responsibility system.

An experiment is underway in Shunyi County in the Beijing Municipality that approximates a return to many of the features of the commune system. This is a county in which there has been rapid development of non-agricultural employment. The total labor force of the county is 220,000 and has been at that level during the 1980s. In 1980, only 20,000 were employed in township industries; in 1988 there were 17,000 plus 20,000 contract workers from outside. County enterprises have not expanded and employ 10,000. While the majority of workers were agricultural in 1980, only 28,000 are now engaged in crop farming. The county followed the household responsibility system through 1986 with seeming success: per capita income rose from 140 *yuan* in 1978 to 859 in 1984 and grain production increased by 6.4 percent annually from 1978 to 1984. But growth slowed to about 1 percent for the next two years. At the end of the household responsibility system phase, farmers were receiving most of their income from tertiary and secondary occupations; most had become part-time farmers and, as it was said, "did not have high incentive for grain production."

The reform consisted of establishing "cooperative farms" of 3,000 *mu* (500 acres). The farms do not own their machinery but are served by tractor stations. The process of change from the household responsibility system to cooperative farms is well illustrated by what happened in one township. In 1987 there were 997 agricultural workers on 35 farms. In August 1988, a new policy reduced the number of farms to 3 and the number of farm workers to 507. Many of the workers who left still had a year or more on their land use rights. How were these farmers induced to give up their rights? We were told that the farm managers employ team leaders, that the team leaders employ people, and that they retain the best people. The other workers were promised jobs in the township industries. We were told that two principles were applied in the transition to the new policies. The first was that both land and enterprises are collectively owned; therefore, one can't occupy a job in both areas. The second was that the minority must obey the majority. What may in fact have been true was that the majority had to obey a very small minority.

The agricultural organization of one township indicates the degree to which the old leadership has regained its authority. A township farm committee includes one representative (the village head) from each of the 16 natural villages. Each of 5 farms in the township has operational teams responsible for 1,000 *mu*. In every case, the team leaders were former brigade leaders. Each team leader supervises from 20 to 30 workers who operate under a contract that relates compensation primarily to output level with little apparent recognition of costs.

It appears that, in an effort to make this experiment a success, agriculture is heavily subsidized by the nonagricultural enterprises. Investment in agriculture (not counting the value of land) is outrageously high: about $1,000 per acre at the official rate of exchange or about $600 at a more realistic rate. The $600 is approximately the same as the investment in buildings, machinery, and livestock, plus half the value of land of investment in Iowa! Given the large difference in labor earnings, it seems obvious that either there is too much investment in Shunyi County or too little in Iowa.

The Secretary of the Communist Party in this county had close contact with national leaders. In addition to his personal traits, his county has two important assets: a golf course used by national leaders, and (reputedly) the best tailor shop in the Beijing Municipality. Thus, the county receives frequent visits from national leaders who are kept informed of the miraculous but undocumented results of the experiment.

Concluding Comments

I do not know what wave the future may bring with respect to the organization of Chinese agriculture. But I fear that the household responsibility system's unpopularity with numerous local leaders is inciting them to regain lost power, authority, and incomes. In any case, as the title of this chapter suggests, noneconomic factors will play a very large role in the path that is followed.

In farming areas where nonagricultural enterprises are profitable and expanding, economic effectiveness may have little to do with how agriculture is organized in the future. So long as industrial reforms falter and fail to provide significant reductions in real prices, the often technologically backward rural nonfarm enterprises can remain very profitable. A further reason for their profitability may well be the discipline of the labor force. There is no such thing as a life-time contract—an iron rice bowl—in the rural enterprises.

Finally, much of the recent criticism of the rural reforms has resulted not from their inherent failure, but from the ineptitude of national policy makers in controlling investments and the money supply. Unnecessary inflation has done enormous harm to the reform process. It has also made it politically impossible to achieve the price reforms prerequisite to any significant decentralization of the economy. The prices of industrial products, as well as some farm products, are especially out of line. The co-existence of inflation and outmoded price relationships supports the continued control of the bureaucrats over the economy.

Notes

1. His results indicated that the complete adoption of the responsibility system would increase agricultural output by 14 percent over and above all the other changes (Lin, 1986, p. 101). In a revision and extension of his work in the dissertation, Lin estimated that the responsibility system was the source of 62 percent of the output growth from 1980 to 1983 (Lin, 1987, p. 25). The full adoption of the responsibility system increased output by 97.7 percent.

2. Their estimate of the incentive effect of the household responsibility system was a residual derived from an estimate of the effect of price increase upon productivity growth. While the theoretical foundation seems appropriate, the estimate of the price effect rested entirely upon the observation of one year's data; namely, price and productivity changes from 1978 and 1979, when there were no institutional changes but a major price increase (McMillan, p. 16).

3. The estimates were made by James Wen.

4. The Big Family was recognized as an acceptable form of organization in Document No. 1, *Circular of the Central Committee of the Communist Party of China Covering the Rural Work in 1984* (January 1, 1984).

References

Central Committee of the Communist Party of China. 1986. *On Arrangements for Rural Work in 1986 by the Central Committee of the CPC and the State Council of the PRC*, Document No.1, January 1. Beijing, International Liason Department, The Research Centre for Rural Development of the State Council of the PRC.

Johnson, D. Gale. 1987. "Agricultural Policy Reforms in the USSR and the People's Republic of China," Office of Agricultural Economics Research, The University of Chicago, Paper No. 87:28, December 2.

Lin, Justin Yifu. 1986. "The Household Responsibility System in China's Agricultural Reform: A Study of the Causes and Effects of An Institutional Change," Ph.D. Dissertation, Department of Economics, The University of Chicago, June.

______ . 1987. "Household Farm, Cooperative Farm, and Efficiency: Evidence from Rural De-Collectivization in China," Economic Growth Center, Yale University, March.

McMillan, John, John Whalley, and Li Jing Zu. 1989. "The Impact of China's Economic Reforms on Agricultural Productivity Growth." *Journal of Political Economy*, Vol. 97, p. 781-807.

State Statistical Bureau of the People's Republic of China. 1989. *China Statistical Yearbook 1988*. English Editor. Hong Kong, International Centre for the Advancement of Science and Technology, Ltd.

Wen, Guanzhong James. "The Current Land Tenure System and Its Impact on the Long Term Performance of Farming Sector: The Case of Modern China," Department of Economics, The University of Chicago.

Yuan, Mu. 1989. News Report on Beijing Domestic Radio Service, 13 January 1989. Translation in FBIS-Chin-89-010 National Service, 17 January, p.42.

Zhao, Huazhou. 1988. "Freedom from 4 Years of Zero Growth—Roundup of the National Rural Work Conference." Translated in FHIS-CHI-88-235, National Affairs, 7 December.

PART TWO

Institutions

Vegetable marketing in Shenyang, Liaoning, China. *(Courtesy Frederick Crook)*

Introduction to Part Two

Francis C. Tuan

Institutional changes and a cohesive rural organization structure often play critical roles for agricultural development in a country, although such technological factors as use of chemical fertilizer and insecticides, better crop varieties and animal breeds, more efficient machinery, and improved irrigation and drainage systems usually receive greater emphasis. Historical experience tells us that, without an integrated governmental and rural organizational setup and appropriate rural institutional changes, reforms of agricultural policies and related economic development plans cannot be effectively implemented. This has been true also in both parts of China during the last 40 years.

In this part of the book, development experiences with an integrated agriculture organizational structure, as established in Taiwan, and revolutionary institutional changes of the farming system, as carried out in Mainland China's fast growing agriculture in the early 1980s, provide two excellent examples to the rest of the world of the pivotal role of effective organizational setup and institutional change during the process of agricultural development.

Mainland China's rapid agricultural growth in the last 10 years, particularly the early 1980s, stemmed from a fundamental institutional change, namely, the establishment of the household responsibility system. This farm household-based production system replaced the commune system, in which production teams comprised of farm households were the basic production units. As a result of this institutional change, farm households, the new basic production units, determine major farming decisions. Consequently, farm families now can initiate production activities, make marketing decisions, and allocate resources to raise output, reduce costs, and maximize income. Farm households responded positively to the institutional changes by applying more inputs, adopting new varieties, breeds, and new technology. As a result, farm output and household income have risen sharply in the past decade.

Justin Lin, formerly with the Research Centre for Rural Development of the State Council and now with the Development Research Centre of the

State Council in Mainland China, is the author of Chapter 4, titled "Technological Choice in Chinese Agriculture." His chapter analyzes the impacts of institutional change upon technological choice in the economy of Mainland China, where market exchanges in land and labor have generally not been allowed. The empirical data employed in his study to test a set of hypotheses demonstrate that the pattern of technological choice in a nonmarket economy such as in Mainland China can be virtually identical to that in market-oriented economies. The reason for such a similarity arises from the fact that the increasing scarcity of a factor raises the marginal productivity of that factor. The income maximizing motive will thus induce decision makers in both market-oriented and socialist economies to search for the kind of technology that facilitates the substitution of the increasingly scarce factor.

Lin further concludes that the institutional change from the communes to a household-based system in the early 1980s created positive incentives to adopt such modern farming inputs as fertilizers and tractors. However, the institutional changes also resulted in a negative demand shift because of their effects on factors affecting supply of the technology. Lin argues that this occurred because the distribution, extension, and dissemination channels of technology under the commune system had been fairly sophisticated and efficient. Indeed, Liu Zhicheng's comments on Chapter 5 (by Mao Yu-Kang) confirm that China does seem to possess a relatively sophisticated extension and dissemination system. Lin concludes that, therefore, the net impact on the demand for tractors and fertilizers after the institutional changes depends upon the relative magnitudes of the positive incentive effect and the negative supply effect.

Mao Yu-Kang, of the Council of Agriculture, Executive Yuan of the Taiwan Government, titled his chapter "Institutional Organizations in Taiwan's Agricultural Development." In it he emphasizes that many factors contributed significantly to the success of rural development in Taiwan, such as land reform, adaptive agricultural research, support services delivery system, rural infrastructure, rural industrialization, etc. Thus, a complementary and cohesive set of *institutional* organizations along with the development of agricultural *technology* suitable to the underlying resource endowment have facilitated Taiwan's rural and agricultural development.

Mao focuses his presentation mainly upon the organizations that have played important roles in agricultural development and describes how all these organizations functioned together to carry out rural economic development plans in Taiwan. He groups all central and local agricultural organizations in Taiwan into three categories: (1) the Sino-American Joint Commission on Rural Reconstruction (JCRR), later changed to the Council of Agriculture, (2) the farmers' organizations, and (3) agricultural research institutions.

Mao's chapter discusses in detail the structure and functions of each individual organization listed under the three categories. More importantly, it points to important characteristics of each of the organizations, as well as to how all these related organizations have functioned together during the last 40 years, in planning and implementing agricultural development policies in Taiwan. In his conclusion, Mao suggests that Taiwan's experience in setting up the agriculture organizational structure may be useful and transferable, to a certain extent, to other economies such as Mainland China, in the course of their rural and agricultural development process.

4

Technological Choice in Chinese Agriculture

Justin Yifu Lin

Introduction

The purposes of this chapter are twofold: to portray the mechanism of technological choice in China, where land and labor markets are severely constrained; and to describe how recent changes in farming institutions from the collective system to the household-based system affect that choice.

Technological change is among the most important factors underlying agricultural productivity growth. This growth can be achieved by two different types of technology: "labor-saving" and "land-saving." The former is designed to facilitate the substitution of power and machinery for labor, the latter to facilitate the substitution of labor or industrial input for land (Hayami and Ruttan, 1985, Chapter 4). An efficient choice between these two categories of technology is critical to achieving rapid productivity growth. It is a well-established hypothesis that farmers in a market economy will be induced, by the changes in factor prices, to search for the technological alternatives that substitute for the increasingly scarce factor of production (Hicks, 1932; Hayami and Ruttan, 1985; and Binswanger and Ruttan, 1978). The Hicks-Hayami-Ruttan hypothesis implies that differences in the relative abundance of factor endowments in an economy will result in differences in the efficient path of technological change.

As indicated in Table 4.1, China is a vast country with dramatic disparities in land and labor endowments from South to North, and from East to West. Efficient technological choice in China requires taking each region's different factor endowments into consideration. However, market exchanges of land and labor between production units were prohibited in the collective system, and severely limited even after the reinstitution of the household-based farming system (Lin, 1989a). Therefore, the relative

TABLE 4.1 Land and Labor Endowments in Each Province

Province	*Labor*[a]	*Cultivated Land*[b]	*Land-Labor Ratio*[c]	*% of Area Irrigated*[d]	*Multiple Cropping Index*[e]	*Effective Land-Labor Ratio*[f]
Guizhou	10,087	28,480	2.8	24	153	3.8
Sichuan	38,712	98,109	2.5	47	181	3.9
Zhejiang	14,030	27,249	1.9	84	252	4.0
Yunnan	12,808	42,488	3.3	34	140	4.3
Guangdong	19,061	47,130	2.5	65	200	4.4
Guangxi	13,963	39,301	2.8	54	177	4.4
Fujiang	7,083	19,240	2.7	64	189	4.5
Tibet	825	3,437	4.2	53	93	4.6
Hunan	20,239	50,998	2.5	82	218	4.8
Henan	25,370	106,508	4.2	45	160	4.9
Shanghai	2,071	5,249	2.5	98	218	4.9
Anhui	17,478	66,518	3.8	50	177	5.9
Jiangsu	20,068	69,451	3.5	75	184	5.9
Shandong	24,988	107,728	4.3	63	146	6.1
Hubei	14,572	55,481	3.8	63	200	6.6
Tianjin	1,391	6,879	4.9	67	133	6.7
Beijing	1,372	6,343	4.6	81	151	6.9
Hebei	17,481	98,551	5.6	54	131	7.3
Jiangxi	9,305	35,753	3.8	74	229	7.4
Shaanxi	9,124	56,377	6.2	33	127	7.6
Qinhai	1,104	8,640	7.8	27	87	7.7

(Continues)

TABLE 4.1 *(continued)*

Province	*Labor*[a]	*Cultivated Land*[b]	*Land-Labor Ratio*[c]	*% of Area Irrigated*[d]	*Multiple Cropping Index*[e]	*Effective Land-Labor Ratio*[f]
Liaoning	6,488	54,814	8.4	19	102	8.9
Gansu	5,841	53,425	9.1	24	98	9.5
Shanxi	6,647	58,076	8.7	28	107	9.6
Ningxia	1,061	12,605	11.9	28	101	12.8
I. Mongolia	4,657	75,974	16.3	20	91	16.3
Jilin	3,845	60,895	15.6	18	100	16.3
Xinjiang	2,570	47,422	18.5	83	92	21.4
Heilongjiang	4,110	131,273	31.9	7	98	32.1

[a]Agricultural labor force excluding workers in village-run industry, unit = 1,000 workers.
[b]Cultivated land unit = 1,000 mu.
[c]Col. 2/col. 1.
[d]% of cultivated land irrigated.
[e]Unit = %.
[f]Effective land-labor ratio is the land-labor ratio adjusted for irrigation and multiple cropping; its formula is: effective land-labor ratio = Land-Labor ratio x (1 + % of area irrigated/4) x [1 + (Multiple cropping index - 100)/2]. See A. M. Tang, *An Analytical and Empirical Investigation of Agriculture in Mainland China 1952-1980* (Seattle: University of Washington Press,1984) for the rationale of these adjustments.

Source: China Agriculture Yearbook 1984.

scarcity of a factor cannot be reflected by its relative price. It is thus interesting to ask, "What are the implications of these factor market constraints for the path of technological changes?" and "What are the impacts of farming institutional reforms upon technological choice?"

I will argue that, although the inducement mechanism differs, the Hicks-Hayami-Ruttan hypothesis is also applicable to an economy with no factor markets. While the relative scarcity of a factor is reflected by relative prices in a market economy, the relative scarcity of a factor in a nonmarket economy is reflected by its relative marginal productivity. Increasing scarcity raises the marginal productivity of a factor. A decision maker in a nonmarket economy is thus induced to search for the type of technology that substitutes for the factor with an increasingly large marginal productivity. Consequently, the efficient path of technological choice in a nonmarket economy is similar to that in a market economy. This theoretical conclusion should hold for both collective and household-based farming systems.

This chapter will be organized as follows. I will first briefly review the history of technological development and institutional changes in Chinese agriculture. A model of technological choice in an economy characterized by the absence of land and labor markets is next presented. Several testable hypotheses are derived in this section. After a short description of the data set, the method of empirical testing and the results are then discussed and presented. Some concluding remarks are provided in the last section.

Technological and Institutional Changes in Chinese Agriculture

In China, the changes in farming institutions and the choice of agricultural technology have been strongly influenced by the development strategy in the industrial sector. Upon the founding of the People's Republic in 1949, the government inherited a war-torn economy with about 90 percent of the population living in rural areas.[1] As an advanced capital-intensive industry was thought to be both the main characteristic of a developed economy and the quickest path to enhanced national power, China adopted a Stalin-type heavy-industry-oriented development strategy as soon as the economy recovered from the war destruction.[2]

The heavy-industry-oriented development strategy resulted in a great demand for agricultural products.[3] Consequently, agricultural stagnation and poor harvests not only affect food supply but also have a direct negative impact upon industrial expansion.[4] As the government was reluctant to divert investment resources from industry to agriculture, it opted for a new agricultural development strategy that would permit and foster the simultaneous development of agriculture and industry. The core of this strategy involved mass mobilization of rural labor to work on such labor-intensive investment projects as irrigation, flood control, and land reclamation; and an effort to raise unit yields in agriculture through such traditional methods and inputs as closer planting, more careful weeding, and more organic fertilizer. The collectivization of agriculture was the strategy that the Chinese government believed would perform these functions.[5]

The official approach to collectivization was initially cautious. Peasants were encouraged and induced to join various forms of cooperatives on a voluntary basis. The movement was surprisingly successful in the initial stages. Encountering no active resistance from the peasantry, it was carried out relatively smoothly. Encouraged by this experience, government leaders in 1958 forced upon the rural sector the broader approach of the people's commune with an average size of 5,000 households.[6] Payment to commune members was mainly according to subsistence needs and partly according to work performed.

Although, as expected, billions of man-days of labor were mobilized, the communal movement ended in a profound agricultural crisis between 1959 and 1961. This crisis was estimated to have resulted in 27 million deaths during these three disastrous years (Lin, 1989c).[7] Although the communes were not abolished after the crisis, agricultural operations and management were delegated, starting from 1962, to a much smaller unit, called the production team, which consisted of about 20 to 30 neighboring households. In this new system, land was jointly owned by the commune, the brigade, and the production team. Remuneration, based on the work points earned by each member, was made within the production team. This system was not changed until the household-based farming system reform inaugurated in 1979.

After the crisis, more emphasis was given to modern inputs.[8] As shown in Table 4.2, the use of chemical fertilizer increased rapidly, at a compound rate of 17.9 percent per year between 1962 and 1978. By the 1980s, China had become the third largest consumer and producer of chemical fertilizer in the world (Stone, 1988; Stone, 1986). Similarly, the pace of mechanization also accelerated after 1962, especially during the 1970s. Measured in total horsepower, the average annual growth rate between 1962 and 1978 was 18.6 percent.[9] Meanwhile, the government developed a sophisticated system of agricultural research, breeding, adaptation, seed production, dissemination and extension. By the early 1970s, almost all the traditional varieties of rice and wheat had been replaced by modern dwarf varieties introduced during the 1960s. Modern varieties of corn, cotton, and other crops were also introduced and promoted in the 1960s and 1970s (Zhu Rong, 1988). Hybrid rice started to replace dwarf varieties of rice in 1976 and rose to 28 percent of rice area by 1986. The innovation and commercialization of hybrid rice in China were considered the most important achievement in rice breeding in the 1970s (Barker and Herdt, 1985).

Although the research and dissemination of modern technology in the collective system were impressive, productivity growth remained moderate at best.[10] The major drawback of the collective system is its incentive structure. Work points were supposed to be accredited according to the amount of effort contributed by each farmer. However, because of difficulties in monitoring agricultural operations, farmers in general received flat work points for each day's work. Due to lack of incentive, a collective system proved less efficient than a household-based farming system (McMillan, Whalley, and Zhu, 1989; Lin, 1987).

After the chaos of the Cultural Revolution, the government started to reconsider rural development policies. Starting in 1978, the authorities implemented a series of major changes, including diversification of the rural economy, production specialization, crop selection in accordance with regional comparative advantage, expansion of rural fairs, and a marked rise in state procurement prices. The original intention was to improve the performance of the collective system. However, a small number of production teams, first secretly, and later with the blessing of local authorities, began to experiment with contracting land and other factors of production to individual households toward the end of 1978. A year later, these teams brought in yields far larger than other teams in the same region. Observing this remarkable effect, the pragmatic leaders in the central government endorsed this new form of household-based farming system. By the end of 1983, over 97 percent of production teams in China had converted to the new system, in which land was leased to individual households for 15 years. Farmers, after fulfilling their quota obligations, could keep the remainder of their harvest for sale on the market. In addition to the institutional changes of the last decade, the pace of mechanization and the consumption of chemical fertilizer were also accelerated during the 1980s (Table 4.2).

A Model of Technological Choice in a Nonmarket Economy

The technological development in Mainland China in the last four decades has been remarkable. However, this chapter questions how responsive the technological development in China has been in reflecting its relative scarcity of factor endowments, and how institutional changes have affected the behavior of technological choice.

In addition to farming institutions in general, rural institutions in China are different from those in other developing economies in many ways. Land is collectively owned. Market exchanges in land between different production units in the collective system were soon outlawed after 1949. Leasing out land for rent was also prohibited, because rent was taken as a form of exploitation in Marxian ideology. Labor market exchange between different production units in the collective system was prohibited for the same reason. The household-based farming system reform in the early 1980s brought some relaxation of this ideological rigidity. A household was

TABLE 4.2 Indices of Labor, Land, Tractors, Fertilizer, and Institutions

Year	*Labor (1952=100)*	*Land (1952=100)*	*Tractor (1957=100)*	*Chemical Fertilizer (1965=100)*	*Household Farm*[a] *%*
1952	100	100	10	0	67
1957	101	104	100	0	0
1962	85	95	325	0	0
1965	92	96	244	100	0
1970	110	94	608	178	0
1971	112	93	1,637	204	0
1972	112	92	2,345	235	0
1973	114	93	3,126	287	0
1974	114	92	3,909	270	0
1975	112	92	4,823	298	0
1976	112	92	5,817	324	0
1977	111	92	7,095	358	0
1978	111	92	8,574	490	0
1979	114	92	8,857	590	1
1980	116	92	11,661	658	14
1981	120	91	12,541	694	45
1982	121	91	13,470	765	80
1983	123	91	14,932	829	98
1984	123	91	15,248	841	99
1985	118	90	17,776	821	99
1986	119	89	19,656	898	99
1987	121	89	22,274	942	99

[a]Percentage of households in the household-based farming institution.

Source: State Statistical Bureau.

allowed to release operated land to other households in case the original household migrated to other areas or jobs.[11] A farm household was allowed to hire workers, not exceeding a certain ceiling. Even today, however, land and labor markets exist only marginally, except in the vicinities of a few major cities. The purpose of this section is to build a behavioral model of technology choice within such an institutional framework.

For the purpose of simplicity, at any given time (t), a farm in an area (i) is assumed to produce only one crop with two endowed factors, land (K) and labor (L). Since no factor markets exist, it is impossible to alter the endowments through factor market transactions. However, there exist two types of technology: land-saving (k) and labor-saving (l). The costs of adopting these two technologies are assumed identical. For any given production cycle, a farm is assumed, without loss of generality, to be endowed with a unit of budget to purchase an appropriate combination of these two technologies to maximize the output. The maximization problem for a farm in area i at time t can be expressed as follows:

$$(1)\ \max_{l_{i,t},\,k_{i,t}} \quad Y_{i,t}(l_{i,t}\, k_{i,t} \mid L_{i,t}\; K_{i,t}, I_{i,t}\; C_i, t)$$
$$= Y\,[L_{i,t} + l_{i,t}, K_{i,t} + k_{i,t} \mid I_{i,t}, C_i, t],$$
$$\text{subject to } l_{i,t} + k_{i,t} = 1,\ l_{i,t} \geq 0,\ k_{i,t} \geq 0,$$

where Y(.) is a well-behaved production function; subscripts i and t indicate area and time; I indicates the type of farming institution (collective or household farm); C is a vector of region-specific, time-invariant characteristics, such as soil type, and rainfall, which affect the marginal productivity of l and k; and t is included to indicate the region-invariant time-dependent changes in the demand shifts that may arise from the changes in the prices and/or technical efficiency of these two technologies. The technological choice is made by a production team leader in the collective system and by a household head in the household-based farming system. By substitution, Equation (1) can be expressed as:

$$(2)\ \max_{l_{i,t}} \quad Y_{i,t}(l_{i,t} \mid L_{i,t}, K_{i,t}\; I_{i,t}, C_i, t)$$
$$= Y\,[L_{i,t} + l_{i,t}, K_{i,t} + 1 - l_{i,t} \mid I_{i,t}, C_i, t].$$

The first order condition for optimality requires that

$$(3)\ Y_1 - Y_2 = 0,$$

where Y_1 and Y_2 are, respectively, the partial derivatives of Y with respect to land and labor. Equation (3) implies that, at the margin for optimality to hold, a decision maker equates the gains from adopting the labor-saving technology with the gains from adopting the land-saving technology.

The efficient path of technological choice in response to the difference in land and labor endowments is implicitly defined in equation (3). From the implicit function theorem, we can obtain the following relations:

$$(4)\ d1_{i,t} / dL_{i,t} = -[Y_{11} - Y_{21}] / \text{s.o.c.} < 0,$$

$$(5)\ d1_{i,t} / dK_{i,t} = -[Y_{12} - Y_{22}] / \text{s.o.c.} > 0.$$

$$(6)\ dk_{i,t} / dL_{i,t} = -d1_{i,t}/dL_{i,t} > 0,$$

$$(7)\ dk_{i,t} / dK_{i,t} = -d1_{i,t} / dT_{i,t} < 0,$$

where s.o.c. stands for the second order condition. According to the assumptions of the production function, s.o.c. < O.

The implications of equations (4), (5), (6), and (7) are intuitively correct. Given the level of land endowments, a farm will demand less labor-saving technology and more land-saving technology if the labor endowment in that farm increases. Likewise, given the level of labor endowment, a farm will demand less land-saving technology and more labor-saving technology if the land endowment in that farm increases. These results hold for both time series and cross-sectional comparisons. The demand functions for the land-saving technology and labor-saving technology can thus be expressed as follows:

$$(8)\ 1_{i,t} = 1(\underset{(-)}{L_{i,t}}, \underset{(+)}{K_{i,t}} \mid I_{i,t}, C_i\ t).$$

$$(9)\ k_{i,t} = k(\underset{(+)}{L_{i,t}}, \underset{(-)}{K_{i,t}}, C_i, t).$$

The signs within the parentheses indicate the direction of impact when the respective factor changes. From equations (8) and (9) we conclude that, in the absence of distortions other than the prohibition of exchanges in land and labor markets, the patterns of technology choice in a socialist economy shall mirror Hicks-Hayami-Ruttan predictions for a market economy. The reason for this similarity is as follows. The increasing scarcity of a factor will raise its marginal productivity; therefore, income maximization will induce farmers to seek technologies that facilitate the substitution of the increasingly scarce factor of production. Equations (8) and (9) also indicate that the institutional change from a collective to a household-based system will not alter the efficient path of technology choice, although the *magnitude* of response to the relative scarcity of land and labor endowments may differ.

The preceding theoretical analysis can be summarized into three empirically testable hypotheses:

Hypothesis I: The increasing scarcity of land will induce farmers in a socialist economy to search for more land-saving and less labor-saving technology.

Hypothesis II: The increasing scarcity of labor will induce farmers in a socialist economy to search for more labor-saving and less land-saving technology.

Hypothesis III: The change in farming institution from the collective system to the household-based system will not alter the conclusions of the above two hypotheses.

Data

To test the hypotheses derived in the preceding section, we need to have data on labor-saving and land-saving technologies, as well as information on land endowment, labor endowment, and farming institutions. Among conventionally used agricultural inputs, tractors are typically identified with labor-saving technology and chemical fertilizers, with land-saving technology. The empirical tests will thus include time-series data from 1970 to 1987 on land, labor, tractors, chemical fertilizers, and farming institutions for 28 of the 29 provinces and municipalities. These data were made available to the author by the State Statistical Bureau.[12]

Land refers to total cultivated land, labor to the total number of farm workers. Tractors are measured by the total horsepower of various types of tractors. Chemical fertilizers refer to the gross weight consumed of fertilizers, including N, P, and K. Data on land, labor, tractors, and chemical fertilizers were not adjusted for differences in quality. The farming institution everywhere was the production team system before 1978 and the household-based system after 1984. For the transition period between 1979 and 1983, detailed data on the percentage of production teams in each province that had converted to the household-based farming system between 1981 to 1983 are also available in the data set. The number of production teams that had converted to the household-based farming system rose from 1.02 percent at the end of 1979 to 14.4 percent by December 1980. Since detailed information for individual provinces is not available for 1979 and 1980, it will be assumed that the farming institution in each province for these two years was the production team system. Summary information on labor, land, tractors, fertilizers, and farming institutions at the national aggregate level is reported in Table 4.2.

Functional Form and Empirical Results

A specific functional form for equations (8) and (9) was estimated to test the hypotheses. For simplicity, the functions were assumed to be linear in land and labor. Since the size of provinces differed, data on tractors, fertilizers, labor, and land were normalized by the number of teams in each province in 1980. For the purpose of estimating the impacts of farming institutional change on technological demand, the coefficients of land and labor were allowed to differ under these two different institutions. In addition, to accommodate the potential impacts on demand due to shifts in technology supply caused by the institutional change, the intercepts of the demand functions were also allowed to differ.[13] Equations (8) and (9) suggest that the demand for technologies may depend upon certain region-specific time-invariant factors, such as soil type and average rainfall; upon certain region-invariant time-dependent factors, such as the change in factor or product prices; and upon the improvement in the technology itself. The demand functions thus also included 27 regional dummies and 17 year dummies to test for the above possibilities. To be specific, the functions estimated were as follows:

$$(10)\quad 1_{i,t} = a_0 + a_{il} I_{i,t} + a_2 L_{i,t} + a_3 I_{i,t} L_{i,t} + a_4 K_{i,t} + a_5 I_{i,t} K_{i,t} + a_6 C_1 + \dots + a_{32} C_{27} + a_{33} Y_{71} + \dots + a_{49} Y_{87} + \mu$$

$$(11)\quad k_{i,t} = b_0 + b_1 I_{i,t} + b_2 L_{i,t} + b_3 I_{i,t} L_{i,t} + b_4 K_{i,t} + b_5 I_{i,t} K_{i,t} + b_6 C_1 + \dots + b_{32} C_{27} + b_{33} Y_{71} + \dots + b_{49} Y_{87} + \nu,$$

where $l_{i,t}$ was the total horsepower of tractors, $k_{i,t}$ was the amount of chemical fertilizers consumed, $I_{i,t}$ was an institutional indicator (percentage of production teams converted to household-based farming system), $L_{i,t}$ and $K_{i,t}$ were respectively labor and land, C_is represented regional dummies, Y_ts were year dummies, and μ and ν were residual terms.[14] It is noteworthy that the number of workers, $L_{i,t}$ in equations (10) and (11), is an endogenous variable. How large a population can be sustained on a given unit of land also depends upon the soil type, climate, rainfall, and other region-specific factors, which were included in the list of omitted variables and are represented by the residual terms, μ and ν. The number of workers in an area is a function of the size of population in that area. Therefore, $L_{i,t}$ shall be correlated with μ and ν in Equations (10) and (11). However, the covariance estimator of the ordinary least-squares model is still a linear unbiased estimator, because equations (10) and (11) were specified in the form fixed-effects model (Hsiao, 1986, p. 72). We thus used the ordinary least-squares model to estimate equations (10) and (11). The results are reported as follows.

$$(12)\ l_{i,t} = \underset{}{5.88} - \underset{(-11.56)}{41.24 I_{i,t}} - \underset{(-9.44)}{.34 L_{i,t}} + \underset{(5.00)}{.25 I_{i,t} L_{i,t}} + \underset{(7.15)}{.03 K_{i,t}} + \underset{(23.28)}{.03 I_{i,t} K_{i,t}} + a_6 e_1 + \ldots + a_{49} Y_{87}$$

$$\bar{R}^2 = .931;$$

$$(13)\ k_{i,t} = .72 - \underset{(-1.64)}{3.38 I_{i,t}} + \underset{(1.27)}{.026 L_{i,t}} - \underset{(-.73)}{.021 I_{i,t} L_{i,t}} - \underset{(-4.27)}{.010 K_{i,t}} + \underset{(2.22)}{.002 I_{i,t} K_{i,t}} + b_6 C_1 \ldots + b_{49} Y_{87}$$

$$\bar{R}^2 = .890.$$

Figures in parentheses are t-statistics. The estimated coefficients for regional dummies and year dummies are suppressed in the expressions. According to equations (12) and (13), the demand functions for chemical fertilizers and tractors under the collective system and household-based farming system can be written separately as follows (the constant, regional dummies, and year dummies are suppressed).

Demand function for tractors:

$$(14)\ l_{i,t} = \underset{(-9.44)^{***}}{-.34 L_{i,t}} \quad \underset{(7.14)^{***}}{+.03 K_{i,t}} \text{ (the collective system)}$$

$$(15)\ l_{i,t} = \underset{(-11.56)^{***}}{-41.24} \quad \underset{(-1.61)^{*}}{.08 L_{i,t}} \quad \underset{(14.2)^{***}}{+.06 K_{i,t}} \text{ (the household-based system)}$$

Demand function for chemical fertilizers:

$$(16)\ k_{i,t} = \underset{(1.27)}{.026 L_{i,t}} \quad \underset{(-4.27)^{***}}{-.010 K_{i,t}} \text{ (the collective system)}$$

$$(17)\ k_{i,t} = \underset{(-1.63)^{*}}{-3.38} \quad \underset{(.15)}{+.005 L_{i,t}} \quad \underset{(-3.57)^{***}}{-.008 K_{i,t}} \text{ (the household-based system)}$$

Note: * and *** respectively indicate significance at the 0.1 and 0.01 level.

From the demand equations (14) through (17), we find that the results are consistent with the implications of Hypotheses I and II. All the estimated coefficients for labor and land had the expected signs and the estimated coefficients were all significantly different from zero except for the coefficients of labor in the chemical fertilizer demand functions. The signs of the coefficients were not altered by the change from the collective system to the household-based system, which is consistent with Hypothesis III.

The estimated coefficients for labor were not significantly different from zero in the fertilizer demand function. This may arise from the fact that chemical fertilizer markets are highly regulated and segmented in China. About three-quarters of fertilizers currently sold in China are tied directly to government quota procurements of grain, cotton, oilseed, etc. Furthermore, little interregional trade of chemical fertilizers exists in China (Stone, 1986). Therefore, the demand in each region is likely constrained by the availability of fertilizers in that region. This explanation is partly supported by the fact that the adjusted R^2 for the fertilizer demand function, equation (13), was lower than that of the tractor demand function, equation (12).

The institutional change from collective to household-based farming, as Hypothesis III postulates and the empirical evidence confirms, will not alter the ways that farmers respond to the relative scarcity of land and labor. The impacts of this institutional change upon total demand for tractors and chemical fertilizers can be analyzed both by how the change affects the incentives and by demand shifts related to changes in factors affecting the supplies of these two factors. The incentives to adopt are measured by the coefficients of land and labor in the demand functions. From the interaction terms in equations (12) and (13), we find that the estimated coefficients of interaction terms all had positive values significantly different from zero, except the coefficient for labor in the fertilizer demand function, which is not significantly different from zero. This implies that the incentives to adopt tractors and fertilizers increased after changing from the collective system to the household-based system. Nevertheless, the institutional change resulted in negative demand shifts due to its effects upon factors affecting supply of these technologies. The negative impacts are measured by the coefficients of $I_{i,t}$ in equations (12) and (13). These negative demand shifts should not be surprising. As Perkins and Yusuf (1984, Chapter 4), Stone (1988), and others have noted, the distribution, extension, and dissemination system of technology under the collective system was very sophisticated and efficient. The collapse of the collective system must have disrupted the network of the original supply system. The total impact on the demand for tractors and fertilizers depends upon the relative magnitudes of the positive incentive effect and the negative supply effect.[15]

Concluding Remarks

This chapter presents a model of technological choice in an economy in which market exchanges in labor and land are prohibited. It is found that, if there are no other distortions, the pattern of choices in such an economy will be virtually identical to that in a market economy. Choice is similar because the increasing scarcity of a factor raises its marginal productivity. The income-maximizing motive will thus induce a decision maker in such an economy to search for technologies that facilitate substitution of the increasingly scarce factor. The rising marginal productivity of a factor carries the same signal as the rising relative price of a factor in a market economy. The empirical evidence from China is consistent with the implications of the model.

This study also suggests a way to analyze the impacts upon technological demand of an institutional change from collective to household-based farming. The impacts were measured both by changes in adoption incentives and by shifts arising from changes in factors affecting the supply of technologies. The empirical evidence shows that such an institutional change in Mainland China had a positive incentive effect on the demand for both tractors and chemical fertilizers. However, the same institutional change resulted in a negative demand shift for both inputs arising from the collapse of the very efficient distribution, dissemination, and extension network under collective farming. The disruption of this supply network increased the costs of acquiring tractors and fertilizers.

Since the pattern of technological choice in a nonmarket economy is not demonstrably different from that in a market economy, patterns of technological innovation will also be similar. Other empirical evidence from Mainland China bears out this statement (Lin, 1989b).

Notes

1. National statistics show that 89.4 percent of the population lived in rural areas, and industry consisted of only 12.6 percent of national income in 1949 (see State Statistical Bureau, 1987, p. 89; and 1987a, p. 11).

2. Wu (1965) argues that the same strategy would have been adopted even if the Communist Party had not been in power. He shows that the symptoms of "take-off" had been exhibited in a number of ways under the National Government's rule in the 1930s. T. N. Srinivasan (1984) also argues that the dominate views among development economists in the early 1950s were supportive of a heavy-industry-oriented strategy for developing countries.

3. First, the urban population increased dramatically from 57.65 million in 1949 to 71.63 million in 1952, and to 99.49 million in 1957 (State Statistical Bureau, 1986,

p. 91). As the heavy-industry development strategy would not allow using large amounts of scarce foreign reserves to import food for urban consumption, meeting the increasing food demand in the urban areas hinged upon the growth of domestic grain production. Second, as the bulk of China's exports had been agricultural and sideline products, especially in the 1950s (see Table 4.2), the country's capacity to import capital goods for industrialization depended upon agriculture's growth. Third, agriculture was the main source of raw material for many industries, such as textile and food processing.

4. This argument is clearly supported by the fact that the heavy-industry-oriented development strategy had to temporarily give way to the "agriculture-first strategy" after the failure in the harvests caused by the collectivization of agriculture in the late 1950s.

5. Of course, the reasons for the collectivization of agriculture were numerous. The desire of the Party to consolidate its control over the countryside and the desire to eliminate the disparity in income distribution in rural areas are just two often-cited arguments. Among the reasons, it is worth mentioning that collectivization of agriculture in 1958 was also treated as a convenient vehicle for securing agricultural products for urban industries and residents (Luo, 1985, p. 53).

6. The main rationale of collectivization was rooted in the notion that mobilizing rural surplus labor would increase rural capital formation and, hence, increase production. However, if a collective farm of 150 households provided a basis for mobilizing labor for work projects within the collective, the collective farm did not solve the problem of mobilizing labor for large projects, such as digging irrigation canals, building dams, or the like. These kinds of projects would in general require the simultaneous participation of labor from several dozens of collective farms. The obvious solution for large-scale labor mobilization was to pool 20 or 30 collective farms of 150 households into a large collective unit. For this reason, from the end of August to the beginning of November 1958 (only two months), 753,000 collective farms were transformed into 26,000 communes, which consisted of 120 million households, over 99 percent of total rural households in China. The average size of a commune was about 5,000 households.

7. The commonly accepted explanations for this crisis were poor weather, bad production management, and incentive issues embedded in the commune system. However, Lin (1989c) argued that the main reason for the sudden production collapse was that the cooperative movement changed from a voluntary to a compulsory movement. Since close supervision is not possible in agricultural production, the success of an agricultural cooperative depends upon a self-enforcing contract in which each member agrees to discipline himself. A self-enforcing contract is sustainable only in a repeated game. The switch from a voluntary to a compulsory cooperative, from the game-theoretical point of view, changed the movement from a repeated game to a one-time game. Therefore, the cooperative was besieged by the Prisoner's Dilemma. As a consequence, agricultural production collapsed.

8. This argument is best demonstrated by the shift in the emphasis of irrigation methods. The irrigated acreage increased 45 percent from 30.5million ha in 1962 to 44.4 million ha in 1987. However, most additional irrigated acreage came from

increased construction of powered tube-wells instead of construction of labor-intensive canals and dams. The share of powered tube-well irrigated acreage in total irrigated acreage increased from 20 percent in 1962 to 56 percent in 1987 (State Statistical Bureau, 1988, p. 233).

9. Mechanization had been one rationale for the collective campaign in the 1950s. The idea survived the great agricultural crisis. In the late 1970s, the complete mechanization of farm operations was pushed even farther forward as the goal of agricultural modernization, which the planners hoped to achieve by 1985.

10. Whether the productivity growth rate in Chinese agriculture before 1978 is positive or negative is debated among scholars. Based on different methods of estimation, some economists argue that the growth rate is positive, while others argue the opposite (see Tang, 1984; Chow, 1985; and Perkins and Yusuf, 1984).

11. In most cases, an out-migration household keeps the entitlement, and gives land to its relatives or friends without obtaining any compensation (Lin, 1989a).

12. The provincial data for 1980 to 1987 on labor, tractors, and chemical fertilizers can also be obtained from various editions of the *China Statistical Yearbook*. The province not included in the data set is Tibet.

13. An institutional change may affect input demand through changes in the costs of supply, and/or through changes in the incentives for adoption. Griliches (1957) suggests that variations in the intercept of a demand function can be identified with the demand shift associated with changes in factors affecting supply; while variations in slope are explained by changes in adoption incentives. Following the suggestion by Griliches, the institutional indicator is included to estimate the impacts of institutional change related to changes in the supplies of tractors and chemical fertilizers, and interaction terms linking this institutional indicator with labor and land are included to estimate the impacts on adoption incentives.

14. The residual terms, μ and v in equations (10) and (11), are correlated. However, there is no gain in using Zeller's Seemingly Unrelated Model to estimate these two equations because the regressors are identical in these two equations (Judge, *et al.*, 1985, Chap. 12).

15. The total number of farm workers was 390 million and total cultivated acreage was 1,438 million *mu* (95.8 million ha) in 1987 (State Statistical Bureau, 1988 p. 212 and p. 224). The total number of production teams in 1980 was 5.66 million (State Statistical Bureau, 1981, p. 131). Therefore, the average size of a team was 69 workers and 254 *mu* (16.9 ha). Based upon the average size and estimated coefficients of institutional indicators and the interaction terms of labor and land in Equations (12) and (13), the net impacts of the institutional changes upon the demand for tractors and fertilizers in 1987 were negative. However, the total demand for fertilizers and tractors increased in 1987 as indicated in Table 4.2. This is because of positive changes associated with the demand factors that were captured by the year dummies. The estimated 1987 year dummies were 47.6 in the tractor demand function and 17.3 in the fertilizer demand function.

References

Barker, Randolph and Robert W. Herdt. 1985. *The Rice Economy of Asia*, Washington, D.C.: Resources for the Future.

Binswanger, Hans P. and Vernon W. Ruttan. 1978. *Induced Innovation: Technology, Institutions, and Development*, Baltimore: John Hopkins University Press.

Chow, Gregory C. 1985. *The Chinese Economy*, New York: Harper & Row.

Hayami, Yujiro, and Vernon W. Ruttan. 1985. *Agricultural Development: An International Perspective*, revised and expanded, Baltimore: Johns Hopkins University Press.

Hicks, John R. 1932. *The Theory of Wages*, London: Macmillan.

Hsiao, Cheng. 1986. *Analysis of Panel Data*, Cambridge: Cambridge University Press.

Judge, George G., W.E. Griffiths, R. Carter Hill, Helmut Lutkepohl, and Tsoung-chao Lee. 1985. *The Theory and Practice of Econometrics*, 2nd edition, New York: Wiley.

Lin, Justin Yifu. 1987. "Household Farm, Cooperative Farm, and Efficiency: Evidence from Rural Decollectivization in China," Center Discussion Paper No. 533, Economic Growth Center, New Haven: Yale University.

_______ . 1989a. "Rural Factor Markets in China after the Household Responsibility Reform," in Bruce Reynolds ed. *Chinese Economic Policy*, New York: Paragon.

_______ . 1989b. "Agricultural Technology Innovation in China: A Study of Public Research Fund Allocation in A Socialist Economy." the Development Institute Working Paper, 89-03, Beijing: The Development Institute.

_______ . 1989c. "Collectivization and China's Agricultural Crisis in 1959-1961," The Development Institute Working Paper, 89-01, Beijing: The Development Institute.

Luo, Hanxian. 1985. *Economic Changes in Rural China*, Beijing: New World Press.

McMillan, John, J. Whalley, and Li Jing Zhu. 1989. "The Impact of China's Economic Reforms on Agricultural Productivity Growth," *Journal of Political Economy*.

Perkins, Dwight and Shahid Yusuf. 1984. *Rural Development in China*, Baltimore: Johns Hopkins University Press.

Srinivasan, T.N. 1984. "Comment on Remembrance of Studies Past: Retracing First Steps by Lord Bauer," in Gerald M. Meier and Dudley Seers eds. *Pioneers in Development*, Oxford: Oxford University Press.

State Statistical Bureau. 1981, 1982, 1983, 1984, 1985, 1986, 1987, 1988. *Zhongguo Tongji Nianjian, 1981-1988* (China Statistical Yearbook), Beijing: China Statistics Press.

State Statistical Bureau. 1987a. *Guominshouru Tongji Ziliao Huibian 1949-1985* (A Compilation of National Income Statistical Data, 1949-1985), Beijing: China Statistical Press.

Stone, Bruce. 1986. "Chinese Fertilizer Application in the 1980s and 1990s: Issues of Growth, Balance, Allocation, Efficiency and Response," in Joint Economic Committee of the U.S. Congress, *China's Economy Looks Toward the Year 2000*, Washington, DC: U.S. Government Printing House, pp. 453-496.

______ . 1988. "Development in Agricultural Technology," in *Special Issue on Food and Agriculture in China During the Post-Mao Period, The China Quarterly*, No.116 (December):767-

Tang, Anthony M. 1984. *An Analytical and Empirical Investigation of Agriculture in Mainland China, 1952-1980*, Taipei: Chung-Hua Institution for Economic Research, (Distributed by University of Washington Press, Seattle).

Wu Yuan-Li. 1965. *The Economy of Communist China: An Introduction*, New York: Praeger.

Zhu, Rong (chief editor). 1988. *Dangdai Zhongguo Nongzuowuye* (Modern China's Agricultural Crops) Modern China Series, Beijing: China Social Sciences Press.

Discussion

Wen-Fu Hsu

I am very pleased to participate in the AAEA-sponsored symposium on Chinese Rural Development, which enables me to have contact with agricultural economists from Mainland China, and to express my personal views on Dr. Lin's chapter titled "Technological Choice in Chinese Agriculture."

The main focus of Dr. Lin's study is to explore how the change in farming institutions affects the adoption of new technology in Chinese agriculture. Change in farming institutions refers to the change from a collective system to a household-based farming system. Dr. Lin takes the use of agricultural machinery and chemical fertilizers as technological factors to substitute, respectively, for the production factors of labor and land. He then measures the relationship between the change in farming institution and the technological choice of farmers. The conclusion drawn from the study is that farmers' technological choices in a nonmarket economy are similar to those in a market economy. In other words, there is no difference in technology choice between the collective system and the household-based farming system.

I have little to add to the structure and methodology used in this paper. Rather, I would like to share several personal viewpoints that might be useful for further discussion.

First of all, it seems to me that the three stated hypotheses are actually general principles of resource use, either in a capitalist economy or a socialist economy. That is to say, if land is a limiting factor, farmers may attempt to use more land-saving technology, such as increased use of chemical fertilizers or improved irrigation facilities, to maximize profit.

Similarly, if labor is a scarce resource, farmers will use more labor-saving technology, such as the adoption of agricultural machinery, to substitute for a part of the labor force.

Differences exist only in terms of whether or not farmers have an incentive to adopt such new technology, or how fast the new technology can be extended, and what is the end result of production growth.

As we know, the change of farming system in Chinese agriculture from a collective to a household-based system, beginning in 1978, has brought substantial production incentives to farmers. This can be seen from statistics that show the percentage of households in the household-based farming system rose to 98 percent within five years (1979-83). At the same time, the application of chemical fertilizers increased 40 percent, and the use of tractors increased 68 percent. It is evident that the household-based farming system was highly favored by farmers.

Moreover, it is the general situation in developing economies that agricultural mechanization cannot easily prevail until industrialization has reached a substantial level. So far as I know, agricultural labor in Mainland China is still abundant, and the wage level is so low that the use of machinery to substitute for labor may not be economical. Under these circumstances, I fail to see why the need for tractors is so urgent in Mainland agriculture. Is it possible that agricultural mechanization is concentrated only in the relatively developed coastal areas? After all, what types of machinery are used there, and how many of them are motorized? These problems are well worth clarifying in the study of the process of economic development.

Finally, the author did not provide further information regarding the impact of changes in the farming system upon agricultural production. From a cursory inspection of the significant increase of tractors and chemical fertilizers (Table 4.2), I think that production must have risen substantially.

5

Institutional Organizations in Taiwan's Agricultural Development

Yu-Kang Mao

Introduction

The economy of Taiwan has grown during the past three decades and continues to grow vigorously. GNP growth in real terms averaged 8.8 percent per year for the period from 1953 to 1987. Per capita national income reached the equivalent of U.S. $4,630 in 1987. These high growth rates were achieved despite rapid population growth and limited natural resources.

Through a series of economic development plans launched since 1953, Taiwan's economy has been transformed from one that was predominantly agricultural, with exports relying heavily on sugar and rice, to one in which a fairly diversified industry is now the leading sector. Table 5.1 shows that in 1953 the share of industry (including mining, manufacturing, electricity, and construction) in net domestic product (NDP) was only 17.7 percent; the share of agriculture was 38.3 percent; and that of the service sector 44 percent. In 1963, the share of industry exceeded that of agriculture for the first time. Since then, the relative importance of agriculture in the national economy has declined continuously. In 1987, the percentage of agriculture in NDP dropped to as low as 6.3 percent.

Taiwan's agriculture is largely made up of about 800,000 farm families with an average farm size of 1.1 ha. The agricultural household population, as a percentage of the total population, declined to 20.7 percent in 1987. Land resources are limited in Taiwan; only a quarter of the total land area or 900,000 ha is arable under the present techniques of production, of which about 60 percent comprises paddy field under irrigation, and the remaining 40 percent dryland.

In the early stages of agricultural development, the expansion of crop production was largely the result of more intensive farming by introducing new and improved techniques, increasing capital inputs like chemical fertilizers, pesticides, and improved seeds, as well as the substantial expansion of multicropping, intercropping, and crop rotation. However, in the last 25 years, crop production has leveled off, and fisheries and livestock have achieved more remarkable results than crops in terms of increased production (see Table 5.2).

As the size of farms in Taiwan has remained very small with no significant change during the past 35 years, it has been difficult for farmers to increase their income from farming. A great proportion of full-time farmers have become part-time farmers or sideline farmers under Taiwan's rapid economic growth. Now the majority of rice farmers in Taiwan are either sideline farmers or part-time farmers. The full-time enterprises are mainly horticultural farms or livestock farms.

Many factors have contributed greatly to the success of agricultural development in Taiwan. These include the land reform, adaptive agricultural research, the support service delivery system, the rural infrastructure, and rural industrialization. However, this chapter will focus only on the organizational aspects of agricultural development. Among the various institutional organizations that have played pivotal roles in agricultural development in Taiwan, the Sino-American Joint Commission on Rural Reconstruction (JCRR), the farmers' organizations, and the agricultural research institutions are often considered the most important. This chapter will discuss these three organizations exclusively.

The Sino-American Joint Commission on Rural Reconstruction (JCRR)

Background and Organization

The JCRR was established on October 1, 1948, in Nanking as a nonpermanent agency for the postwar rural reconstruction of China. With the transfer of JCRR to Taiwan and the reactivation of United States economic aid to the Taiwan Government after the fall of the mainland to the Communists in 1949, the JCRR served as the agricultural arm of the United States mission to Taiwan. This mission was known by a succession of names: ECA, FOA, MSA, ICA, and AID. Whereas the mission devoted its attention mainly to industrial development projects, the JCRR worked on rural and agricultural development. This division of labor between the two agencies, one purely American and the other a bilateral enterprise, was a practical demonstration of how the technicians of two friendly governments could work side by side in a harmonious and fruitful relationship for the attainment of a common objective —the economic development of the less-developed partner.

TABLE 5.1 Changes in the Economic Structure of Taiwan, 1953-1987

Item	*1953*	*1970*	*1987*
Index of GNP at 1981 constant prices	100.0	410.9	1,784.2
Production structure (NDP) (%)			
Agriculture	38.4	18.0	6.3
Industry	17.7	34.5	47.5
Services	43.9	47.5	48.2
Total	100.0	100.0	100.0
Labor structure (%)			
Agriculture	55.6	36.7	15.3
Industry	17.6	28.0	42.7
Services	26.8	35.3	42.0
Total	100.0	100.0	100.0
Foreign trade			
Total Exports (U.S. $ million)	128	1,481	53,538
Total Imports (U.S. $ million)	192	1,524	34,506
Composition of exports (%)			
Agricultural products	13.8	8.6	1.2
Processed agricultural products	77.8	12.8	4.9
Industrial products	8.4	78.6	93.9
Total	100.0	100.0	100.0
Composition of imports (%)			
Agricultural and industrial raw materials	67.1	52.8	62.8
Capital goods	15.6	32.3	29.2
Consumption goods	17.3	4.9	8.0
Total	100.0	100.0	100.0
Per capita national income			
NT $ at 1981 prices	17,863	45,081	142,733
U.S. $ at current prices	159	360	4,630

Source: Council for Economic Planning and Development, *Taiwan Statistical Data Book*, 1988.

TABLE 5.2 Changes in Taiwan's Agriculture, 1953-1987

Item	*1953*	*1970*	*1987*
Cultivated land area (ha)	872,738	905,263	886,281
No. of farm households	702,325	880,274	756,745
Average cultivated land per farm household (ha)	1.24	1.03	1.17
Farm population as % of total population	51.9	40.9	20.7
Agricultural output index	100.0	210.7	289.5
Multiple cropping index	170.4	183.0	142.3
Agricultural production structure (%)			
Crops	71.9	57.2	41.4
Forestry	5.1	5.2	1.2
Fisheries	7.4	13.5	8.2
Livestock	15.6	24.1	29.2
Total	100.0	100.0	100.0
Crop production structure (%)			
Rice	61.3	45.0	26.8
Fruits	2.4	11.8	24.4
Vegetables	3.7	16.0	26.1
Special crops	21.9	14.3	14.4
Other common crops	10.7	12.9	8.3
Total	100.0	100.0	100.0

Source: Council of Agriculture, *Basic Agricultural Statistics*, Republic of China, 1988.

One key characteristic of the JCRR is that it combined the planning function with the actual implementation of research, extension, irrigation, and the revitalization and strengthening of farmers' associations. Responsibility for carrying out these activities was concentrated in only one agency, thus allowing these activities to be implemented in an integrated fashion. It is typical in the developing world for these functions to be conducted by different agencies in an uncoordinated and, therefore, inefficient manner. It is no great exaggeration to speculate that the JCRR was involved in integrating rural development long before the concept had been coined in the literature and espoused by such development agencies as the World Bank. The fact that the same individuals were responsible for planning and actually carrying out agricultural development added an important element of realism to the planning process. Instead of being an academic exercise, planning became a means of conducting a dialogue with the actors (for example, farmers) regarding general objectives and specific targets and the best means of achieving them at the local level.

Another essential characteristic of the JCRR was its effectiveness in operating at the local level in a decentralized fashion, thereby circumventing what might otherwise have been a cumbersome bureaucracy. By involving local farmers in the planning and implementation process, the planning objectives could more easily be translated into specific local actions.

The governing body of the JCRR was the joint commission. Originally, there were five Commissioners, three Chinese and two Americans, with one of the Chinese Commissioners serving as Chairman. They made all the policy decisions and exercised general supervision over the implementation of JCRR-supported projects with the assistance of a technical and an administrative staff. In 1964, by mutual agreement between the Chinese and American governments, the number of commissioners, all Presidential appointees, was reduced to three—two Chinese and one American. The technical divisions of the JCRR were organized on a subject-matter basis. The changing times had necessitated the modifications in the organizational patterns. In 1947, the JCRR had only four technical divisions, namely, the Agricultural Production Increase Division, the Integrated Program Division, the Local Initiative Division, and the Rural Education Division. From 1963 to 1979, the JCRR grew to nine technical divisions and one office: Plant Industry, Animal Industry, Fisheries, Forestry, Irrigation and Engineering, Rural Health, Agricultural Credit, Rural Economics, Farmers' Service, and Planning and Programming. This pattern of evolution showed that the JCRR always adapted its structure to cope with the needs of the times.

Relationship Between the JCRR and Other Government Agencies

The relationship between the JCRR and other Chinese government agencies is not clearly set forth in any document. Being a unique organization, the JCRR's actual relationship with other Chinese institutions was rather *ad hoc* and depended to a large extent on human factors. The JCRR Chairmanship was generally considered to be of ministerial rank. After 1953, when a series of Four-Year Economic Development Plans were implemented, the JCRR developed many functional relations with government organizations in planning and implementing agricultural programs.

The relationship of the JCRR with the agricultural agencies of the provincial government has been amicable since the mainland days, because all parties work for the same objective. The JCRR renders technical and financial assistance to projects of the provincial government agencies. For example, the Department of Agriculture and Forestry of the Taiwan Provincial Government, which carries the burden of agricultural field work in Taiwan, is one of the major recipients of JCRR aid funds. Thus, the relationship between the Department and JCRR has always been most cordial. In fact, three of the JCRR's key staff members served as commissioners of the Department.

From the very beginning, it was understood that the JCRR could work with any level or unit of a sponsoring organization in carrying out its projects. In the JCRR's early years, it preferred to make grants or loans to the lowest unit of an organization so that this unit could be held directly responsible for implementing the project and accounting for funds expended. In this way, bureaucratic control was bypassed, red tape reduced to a minimum, and assistance extended to the end-users at the grass roots. This system made it possible for government agencies to be of real service to the people. Although some Chinese and American observers criticized the Joint Commission's preference for working with so many low-level agencies, the results demonstrated the essential soundness of such an approach.

Program Operation

To apply practical solutions to rural problems, and to realize its policies, the JCRR often had to operate a great number of projects, including crop improvement and irrigation, land reform, farmers' organization, animal husbandry, forestry, rural health, fisheries, agricultural credit, and rural economics. Benefiting from experience, the Joint Commission had developed a program procedure that was spelled out in great detail in an Administrative Order.

A set of criteria relating to programs and procedures was worked out. The four basic preconditions that have guided the JCRR operations throughout the last 30 years are:

1. A need for JCRR services on the part of farmers, who are the direct beneficiaries of the project.
2. A fair distribution of the benefits derived from JCRR projects.
3. A sponsoring agency qualified to undertake any given JCRR project and utilize JCRR assistance effectively.
4. A demonstration of the feasibility of any particular JCRR project before undertaking its broad expansion.

JCRR projects were generally initiated by local agencies that requested technical or financial assistance. Some projects of an innovative nature were initiated by JCRR specialists, who enlisted appropriate agencies as project sponsors. The requests from sponsoring agencies for aid were reviewed by the divisions concerned. In reviewing such requests, the divisions took into consideration the following factors: (1) conformity with approved program objectives, (2) technical feasibility, (3) reliability of cost estimates, (4) availability of funds, and (5) availability of technical and administrative services required. The divisions might then recommend either approval, modification, tabling, or rejection of the requests.

When a project recommended by a particular division had some bearing on, or relation to, another division or divisions, it was passed on to them for comment before it was submitted to the Joint Commission for action. Each project recommended for approval had to include a schedule of payment, which was approved along with the project. However, this approval did not commit the Joint Commission to making payment unless the project progressed satisfactorily. In granting a subsidy for a given project, the JCRR required the sponsoring agency to contribute matching funds commensurate with its financial ability because it was felt that such contributions expressed determination on the part of the sponsor to make the project a success.

The role played by the JCRR in agricultural development was, on the whole, that of a catalytic agent. By providing technical and some financial assistance, the JCRR was able to give direction to agricultural policies, encourage improvements in methods of implementation, motivate projects difficult toactivate, and generate a spirit of self-help among the local agencies. In line with this principle, over 96 percent of all JCRR-supported projects were sponsored by officially recognized agencies and organizations.

It was also the JCRR policy to remit project funds directly to the end-user agency by installments. The disbursement of each installment had to be justified by a "certification of progress" submitted by the Technical Division exercising surveillance over the project and contributing technical assistance to it.

Termination of the JCRR and Establishment of the CAPD

The JCRR made a vital contribution to Taiwan's agricultural development during its 30 years' existence. It furnished a variety of supporting forces in upgrading agriculture in Taiwan. It supplied conceptual leadership in identifying problem areas and in planning remedial measures. It provided support for facilities, equipment, and personnel. It furnished both support and coordination for nationwide research projects, and coordination for research, education, and extension activities. In total, the JCRR provided for a planned and highly integrated agricultural development effort with a high measure of participation by individual farmers.

The accomplishments of the JCRR were attributed to a variety of factors. The most important were a stable political and social order in which projects could operate; a progress-oriented government; a literate, industrious, and thrifty farming population; a steady flow of United States economic aid up to 1965; the flexibility of programs and projects; complete freedom from political pressure or other forms of outside influence; the happy choice of Chinese and American commissioners; and a real concern for the urgent needs of the farmers, whose views were solicited and were taken into consideration in the formulation of policy. Last, but not least, was the spirit of cooperation that had been consistently manifested by Chinese and American staff members.

On March 15, 1979, the JCRR was terminated as a result of the recession of diplomatic relationships between the United States and the Republic of China. The JCRR was then reorganized into the Council for Agricultural Planning and Development (CAPD). Most of the JCRR staff continued to serve in the new organization; some of the JCRR programs were continued by the CAPD, which was the agricultural advisory, planning and coordination agency of the Executive Yuan (Cabinet). On September 20, 1984, the CAPD was again reorganized into the Council of Agriculture. The functions performed by the Council of Agriculture are similar to those of the Ministry of Agriculture in many other governments.

Farmers' Organizations

Technological progress alone is not sufficient to solve the problems of rural areas. Such solutions also call for organizational support. Though academics have paid little attention to the organizational factor, political and administrative leaders in Taiwan have long recognized that farmers' organizations do indeed make a difference. Farmers' organizations will not emerge spontaneously in response to economic policy or technological imperatives, but they are indispensable to the implementation of any development policy.

There are three major organizations of farmers in Taiwan, namely, farmers' associations, irrigation associations, and agricultural cooperatives. Moreover, for fishermen, there are fishermen's associations. Because the organizational setup of the latter is similar to that of farmers' associations, they will not be discussed.

Farmers' Associations

The farmers' association is the most important organization in rural Taiwan. It is a multipurpose cooperative whose main function is to help its members solve their farming, financial, and marketing problems. Therefore, its major services are extension, credit, and marketing. It also serves as a bridge between farmers and the government.

The farmers' association in Taiwan is organized at three levels: provincial, county, and township. At present, there are a total of 303 farmers' associations: 3 provincial/ municipal associations, 14 county/city associations, and 286 township associations. The township association is the key link in the entire system of farmers' associations. It is in direct contact with the farmers and serves them in a variety of ways. In 1988, the total membership of farmers' associations was 1,634,587, of which 63 percent were regular members and 37 percent were associate members. The regular members are *bona fide* farmers who have the right to elect and to be elected to any office of the association. Associate members are those who are not engaged in farming full-time but reside in rural areas. They have no right to vote but can be elected to the post of supervisor.

Major functions of the county farmers' association are to supervise and give technical assistance to the programs run by their constituent township associations. The major functions of the provincial farmers' associations are to assume overall responsibility for supervision, auditing, training, and coordination, and to assist lower-level farmers' associations in the expansion of service activities.

The highest authority of a farmers' association is the members' representatives meeting that is attended by representatives elected by lower levels of farmers' associations (for provincial, municipal, county, or city associations) or small agricultural units (for township or district associations). Separate boards of directors and supervisors are set up by each association. Directors and supervisors are elected by the members' representatives. Policy making is the major function of the board of directors, and auditing is the major function and of the board of supervisors. A general manager is engaged by the board of directors and is responsible to the board for the operation of the association and the management of its business activities.

Six operational sections comprise each township farmers' association: administration, supply and marketing, credit, agricultural extension, accounting, and livestock insurance. A county farmers' association is composed of administrative, economic, extension, accounting, and livestock insurance divisions. The Taiwan Provincial Farmers' Association consists of administrative farm supplies, agricultural marketing, extension, farmer service, accounting, and livestock insurance departments, as well as a Farmers' Association Personnel Training Center.

To better render their various services, the farmers' associations are equipped with such physical facilities as office buildings, warehouses, rice and feed mills, and agricultural processing plants. Among the physical facilities, rice and fertilizer warehouses are the most important. The farmers' associations in Taiwan own a total of 2,650 rice warehouses with a storage capacity of 1,200,000 metric tons of paddy and 735 fertilizer warehouses with a total capacity of 250,000 metric tons. For rice milling, 546 rice mills have been built by the farmers' associations; together they are capable of processing approximately 12,000 metric tons of rice daily.

In Taiwan, agricultural extension services are provided by the farmers' associations in close cooperation with the district agricultural improvement stations. Since 1952, an agricultural extension section with an average of six extension agents has been operating in each township farmers' association. Most of the activities conducted by the extension agents are channelled through farming study groups organized at the farm level. The subjects studied by these groups include crop cultivation, farm machinery, and the application of fertilizers and other agricultural chemicals. The extension work relies mainly upon audio-visual aids, reference materials, training courses, demonstrations, and frequent contact with the farmers through farm and home visits.

Agricultural production has two outstanding characteristics: it is a biological process, and it is handled in small production units. Because of its biological nature, agricultural production is subject to various natural hazards. And because it is done by small farms individually, it cannot enjoy the advantages of large-scale production. These limitations greatly affect farmers' sources and means of obtaining capital, and often turn the farmers to private money lenders. Before the land reform program, landlords were a major source of short-term loans in the rural areas. But such loans usually carried a high interest rate. After the land reform, landlords withdrew from the rural financial market. To fill the gap, the credit section of farmers' associations has played an active role in providing loans for production and marketing. The principle of supervised credit has been applied to some loans. The main source of loanable funds for farmers' associations is from the deposits made by their members. Borrowing from the government and banks makes up for seasonal shortages. Most of the loans are made without collateral but guaranteed by two or more persons. Central Bank interest rates serve as the ceiling for the lending operations of the associations. However, the associations usually extend farm loans at rates below the ceiling.

In each township farmers' association, the supply and marketing section plays an active role in furnishing modern farm inputs and facilitating agricultural marketing. In recent years, the farmers' associations have made considerable efforts to integrate their business vertically. They now operate their own factories to produce feedstuffs and agricultural chemicals. In addition, consumer goods are sold at competitive prices at the farmers' associations. Cooperative marketing of hogs, poultry, and eggs is the most important commercial activity undertaken by the associations for their members. Vegetables and fruits have also become important items of cooperative marketing by the associations in recent years.

In the developed countries where farm production is operated on a larger scale and in a more specialized manner, the farmers may need only a specific kind of service to be performed by a cooperative organization. Therefore, the single-purpose agricultural cooperative is found to be adequate. On the other hand, most farmers in developing countries are smallholders; they require various kinds of assistance that should be provided by an organization in an integrated manner. Under these circumstances, the multipurpose farmers' association has proved to be more useful.

The major functions of a multipurpose farmers' association are to help its members solve their technical, financial, and business problems. The coordinated provision of agricultural extension, farm credit, supply, and marketing services has been considered the most effective solution.

Irrigation Associations

Since paddy rice is the most important food crop in Taiwan, water supply is an essential factor in Taiwan's farm production. A special type of farmers' organization has therefore been formed to operate irrigation facilities and maintain orderly water distribution.

The irrigation association not only handles the use of water resources, it also assists the government in planning and implementing new irrigation projects. There are now a total of 17 irrigation associations organized by farmers according to water regions. The major functions of the irrigation association are: (1) provision and regulation of irrigation water, (2) collection of water fees, and (3) maintenance and construction of irrigation and drainage facilities. Of the 17 irrigation associations, 15 are in Taiwan Province and 2 in Taipei City. In addition, there is a joint irrigation association for coordination of activities of all the irrigation associations.

The chairman of the irrigation association is elected by representatives elected in turn from among the members of the associations. The number of representatives for an association is determined by the size of its irrigation area. The representatives are responsible mainly for the election and recall of chairman, screening of financial reports, approval of operational plans, determination of membership fee rates, and review of matters proposed by members.

The Chairman represents the irrigation association and is assisted by a general manager in performing its functions. A chief engineer is assigned to an association to resolve the technical and engineering matters of the association. In the head office of the association, there are two main divisions, engineering and management, which are directly concerned with irrigation. The former is in charge of the planning, layout, survey, design, and construction of irrigation projects; and the latter is in charge of the operation, management, and maintenance of completed irrigation works. However, in associations serving less than 20,000 ha, a single management division takes care of all technical and engineering matters.

The head office of the association also includes financal and administrative divisions, and accounting, personnel, and security offices. Under the head office of the association, there are working stations that perform most of the tasks of the association in the field. The management division in the head office exercises direct supervision over the irrigation working stations. Each working station handles an area of 1,000-2,000 ha.

Under the working stations, irrigation groups are organized by members of the association as basic units for irrigation and drainage operation and management at the farm level, each handling an area of 50-150 ha. Members of such groups serve gratis, are not association staff, receive only a small allowance for attending the meetings of the association, and are supplied with materials needed for maintenance of the facilities. Each irrigation group consists of several irrigation teams of 10-15 members. The group heads and the team leaders are elected by the members.

In addition to assisting in irrigation operation and management, the irrigation group also contributes labor for the annual maintenance of facilities at the farm level, and helps in planning and setting up common rice nurseries, collecting membership fees, etc. The technical staff of the irrigation association operate and control water delivery at main canals and laterals. From turnout gates through farm ditches to individual fields, the water is distributed by the irrigation groups under the supervision of the association staff.

Usually, the management division in the head office of the association annually prepares a master irrigation plan according to past experiences, government policy, production goals, and water availability for distribution to the local working stations. Then, meetings with the irrigation groups are called to discuss and work out detailed irrigation plans, which are then reported to the head office for approval and compilation.

The water delivery schedule may be revised to meet actual needs when it is put into operation, by taking account of the actual water sources and amounts of effective rainfall during the irrigation period. When water shortage occurs and the available water is not sufficient for regular or normal irrigation, a strict rotational irrigation plan is followed. Records of the irrigation work done are prepared by the working stations and reported to the management division in the head office for checking and for reference in future irrigation planning.

In addition to operation and management, the association is also responsible for the proper maintenance of the irrigation and drainage facilities. This maintenance work is generally divided into three categories, namely, routine maintenance, annual repairs, and emergency repairs. Routine maintenance is designed to make immediate repairs on the spot upon discovery by patrolmen. Annual repairs are made during a planned suspension of irrigation to insure proper operation of the facilities for the next irrigation season. Emergency repairs are mainly made after flood damage. When such damage is heavy, financial aid and technical assistance are required from the government. The costs of routine maintenance and annual repairs are usually paid out of the association's own budget.

Irrigation associations may collect fees from their members. The annual budget for financing their regular operation comes mostly from the collection of ordinary membership fees, for which different rates are set according to water use. In principle, the total amount of the collected membership fees must meet the annual requirements for administrative and personnel expenses as well as irrigation operation and maintenance costs.

However, the rates fixed 30 years ago between an upper limit of 300 kg and a lower limit of 20 kg of paddy per ha per year (collected in cash according to the prevailing price of paddy) are proving inadequate. Due to increasing personnel expenses, labor wages and prices of materials, it has been financially difficult for the associations to fulfill their responsibility in irrigation operation and management. Recently, the government has provided a generous subsidy to the associations for maintenance and improvement of the facilities. Capital investment in irrigation and drainage facilities has also become the sole responsibility of the government. In spite of these financial difficulties, the systems of irrigation and management adopted by the irrigation associations have continued to play an important role, not only in the increase of agricultural production, but also in the improvement of rural infrastructures that enhance the social welfare of rural people.

Fruit-Marketing Cooperative

The Taiwan Provincial Fruit-Marketing Cooperative is a cooperative organization whose purpose is to increase fruit growers' incomes through cooperative marketing. The cooperative has a total of 86,600 members. The main functions of the Cooperative are:

1. To engage in export activities of various fruits from Taiwan. It has been the exclusive banana export agent since 1974, and is the sole supplier of citrus fruits and mangoes to overseas markets.

2. To distribute fruits through cooperative marketing outlets, aiming to increase growers' returns by shortening marketing channels and eliminating unnecessary costs. The products covered are bananas, oranges, pineapple, mangoes, litchi, papaya, grapes, etc.
3. To provide farm supplies and equipment to members to improve farming techniques; such supplies include fertilizers, pesticides, polyethylene bags, and bamboo support-sticks, as well as farm tools and machinery.
4. To develop new markets for Taiwan fruits, especially for such products as *tankan*, mangoes, papaya, and litchi.
5. To carry out research and experimentation to improve marketing practices, and to seek newly improved technology on quality improvement of fruits marketed, such as freshness, shelf life extension, container design, and packing materials.

The pattern of the Fruit-Marketing Cooperative is similar to that of the farmers' association. The highest authority of the Cooperative is the members' representative assembly, which is attended by representatives elected by the members. Supervisors are elected by the members' representatives for a term of one year. Directors are elected directly by members on a regional basis for a term of three years. A president is voted from among the directors themselves as the leader and the representative of the Cooperative. The major function of the board of directors is policy making; that of the board of supervisors is auditing. A general manager is employed by the board of directors to manage the business in accordance with the decisions and policies of the board.

Under the head office, there are eight local branches—Ilan, Taipei, Hsinchu, Taichung, Chianan, Kaohsiung, Pingtung, and Tungtai. These local branches are responsible either for assembling, grading, packaging, and shipping fruits for export, or for transporting them to terminal markets for domestic consumption. A sales office in Japan mainly coordinates business transactions and acts as sales representative of the head office.

In addition, the fruit-marketing cooperative operates a modern container factory producing paper cartons used exclusively for marketing fruits. The cooperative also operates direct supply centers of agricultural produce. These centers are located in Taipei, Taichung and Kaohsiung. The fruits sold at the centers come directly from the member-farmers of the cooperative, and are sold directly to consumers.

In Taiwan, since most farms are small, group action can increase the bargaining power of farmers. With the assistance of the Cooperative, Taiwan's fruit growers export their products successfully to Japan and other countries. A pooling system is employed in making payments to the growers for the fruits marketed cooperatively. In this way, the farmers can avoid exploitation by merchants, reduce costs, and increase farm incomes.

Agricultural Research Institutions

Agricultural research organizations in Taiwan can be classified into four categories: (1) the provincial agricultural research institutes, (2) industrial research institutes, (3) agricultural colleges, and (4) autonomous research organizations. These institutions constitute an island-wide network of agricultural research to provide better agricultural know-how and technology to farmers. Technical innovation has been the prime mover in Taiwan's past agricultural development. Further development of agriculture will depend even more heavily upon the boosting of labor productivity through scientific and technical knowledge developed by intensive research.

Agricultural research institutions are relatively more numerous and more important at the provincial level. Under the Taiwan Provincial Department of Agriculture and Forestry (PDAF), there are six research institutes, namely, the Agricultural Research Institute and its branch stations; the Forestry Research Institute and its field stations; the Livestock Research Institute and its substations and district breeding centers; the Fishery Research Institute and its field stations; the Animal Health Research Institute; and the Agricultural Chemical and Toxic Substances Research Institute. Furthermore, there are six District Agricultural Improvement Stations, the Tea Experiment Station, and the Sericultural Experiment Station.

The industrial research institutes are those sponsored or supported by industries. They include the Sugar Research Institute and its field stations of the Taiwan Sugar Corporation, the Tobacco Research Institute of the Taiwan Tobacco and Wine Monopoly Bureau, the Banana Research Institute of the Taiwan Fruit-Marketing Cooperative, and the Food Industry Research and Development Institute of the Taiwan Food Industry Association.

In university research, there are two colleges of agriculture: National Taiwan University and National Chung-Hsing University. In addition, there are two junior agricultural colleges: Pingtung and Chiayi. The autonomous research organizations set up in Taiwan are the Asian Vegetable Research and Development Center (AVRDC), the Swine Research Center, and the Agricultural Engineering Research Center. The AVRDC is an international research institute with research emphasis on vegetable crops for the benefit of both producers and consumers in the Asian region.

It would be difficult to adequately present all these research institutions in this chapter. However, the Taiwan Provincial Agricultural Research Institute and its Chiayi and Fengshan Stations, the Livestock Research Institute, the Fishery Research Institute, and the District Agricultural Improvement Stations will be briefly introduced due to the important roles they have played in agricultural development of Taiwan.

Taiwan Agricultural Research Institute

The Taiwan Agricultural Research Institute (TARI) was established in 1895 as the provincial agricultural research organization. It was given its present name after World War II. At present, the TARI has six technical departments, namely agronomy, horticulture, agricultural chemistry, applied zoology, plant pathology, and agricultural machinery. It has two branch stations, Chiayi Agricultural Experiment Station and Fengshan Tropical Horticultural Experiment Station, with a total of 210 technical staff members. Research funds are mainly provided by the Provincial Government, the Council of Agriculture, and the National Science Council.

Basic and applied research is undertaken by the Taiwan Agricultural Research Institute in the fields of breeding and genetics, crop physiology, tissue culture, soil fertility, soil chemistry and physics, soil survey, plant nutrition, disease and pest control, toxicology, biological control, and agricultural machinery. In addition, the plant pathology department is responsible for mushroom research, e.g., varietal and cultural improvement, spawn making, and disease and insect studies—all of which contribute to the rapid development of the Taiwan mushroom industry. The crops studied are rice, peanuts, soybeans, corn, and fibers; citrus, pears, peaches, and other fruits; vegetables; flowers and ornamental plants; and other special crops. Significant achievements in recent years include improved rice varieties, single cross hybrid corn, heat-tolerant pears and juicy peaches, sweet persimmons, seedless grapes, waxapple production adjustment, and the biological control of the coconut beetle, red spider mite, and corn borer.

The Chiayi Agricultural Experiment Station has three technical departments: agronomy, horticulture, and plant protection. It undertakes both field and horticultural crop research. It is well known nationally and internationally for sweet potato research. A great number of new varieties of sweet potato have been developed by the application of both conventional and new techniques of crossbreeding. The improvement of rice, especially the development of blast- and plant-hopper resistant varieties, also counts among the main achievements of the station. The horticultural crop research workers have devoted most of their time to improving citrus, passion fruits, bananas, papaya, kiwi, and other tropical fruits of Chinese origin, e.g., litchi and longan. A wide range of tropical fruit trees are maintained in the orchard to develop new varieties of tropical fruit. Pineapple, banana, mango, carambola, passion fruit, and papaya research are conducted by the Fengshan Tropical Horticultural Experiment Station in the extreme tropical region south of Taiwan. The rapid increase of pineapple production in Taiwan after World War II was largely the contribution of this station. The well-known seedless watermelon is also the result of research conducted at this station. A number of improved varieties of cabbage, cauliflower, radish, and other vegetables have also been developed.

Taiwan Livestock Research Institute

The Taiwan Livestock Research Institute (TLRI) was formerly part of TARI and became independent in 1952. It consists of seven departments: animal breeding, animal physiology, forage crops, animal nutrition, milk and meat processing, livestock management, and technical service. In addition, there are three substations in Hsinchu, Hengchun, and Ilan, and six breeding-animal propagation stations at various parts of the island to carry out livestock multiplication and extension. The total technical staff is about 185.

The major research activities are the improvement of cattle, swine, poultry, and forage crops. Native beef cattle were improved by introducing Sindhi blood; recent successes in cattle embryo transfer technology have opened the way to rapid increase of the number of progeny of good breeds needed for genetic improvement. The carcass quality and feed efficiency of swine were improved by the use of three-way crosses and recurrent selection. The growth rate of ducks was much improved by a three-way cross among White Muscovy, Peking Duck, and Common Taiwan Natives; the white feathers, as a result of the cross, are a special feature that meets both domestic and overseas market demand.

Artificial insemination has been practiced on livestock all over the island as a result of the technical accomplishments of this institute. At present, the rates of artificial insemination are 30 percent and 90 percent for hog and cattle, respectively. A semen bank was set up in the TLRI to store both U.S. imported frozen semen and the semen collected from bulls raised in the institute. As a result of the recent worldwide energy crisis, research into the recycling of animal waste to produce methane gas for fuel and rural electrification has yielded good results at the institute. Large-scale use of methane gas in the countryside has been designed and extended.

Taiwan Fishery Research Institute

The Taiwan Fishery Research Institute was established in 1945 for the purpose of improving fishing technology and developing marine resources under the supervision of the Provincial Department of Agriculture and Forestry. This institute is of particular importance because Taiwan is an island on which coastal fishing farms constitute a major part of the agricultural sector. In this institute, there are five departments: marine fishing, aquaculture, fish processing, marine biology, and fishing vessel technology. Seven substations located in Chupei, Lukang, Tainan, Kaohsiung, Taitung, Penghu, and Tungkang carry out various research programs all over the island with a total technical staff of about 100.

Great success has been achieved in recent years in the artificial culture of grey mullet that permitted the fish to be raised in ponds. The new technique was the first in the world and it also made possible the improvement of mullet through breeding. The artificial culture of carp was also a breakthrough. It is now possible to induce spawners to mature in a fish pond and to stimulate them to spawn. Millions of fries are available for domestic production and export to Southeast Asia. Eel culture has also made a great contribution to foreign exchange earnings. In addition, deep-water milk fish culture can increase the yield per hectare by 2-5 times. This technique has also made possible the culture of milk fish in winter. The propagation and culture of fresh water and marine prawn have likewise made a booming business of prawn culture. Water management and the best feeds for both larva and adult prawn are also studied. In addition, hybrid telapia enjoys rapid growth rate and high yield.

District Agricultural Improvement Stations

Under the guidance of the PDAF and working closely with the TARI are six District Agricultural Improvement Stations (DAIS) in the districts of Taoyuan, Taichung, Tainan, Kaohsiung, Taitung, and Hualien. The DAISs are primarily responsible for extension services and experiments on local technical problems.

Each station comprises three technical departments: crop improvement, crop environment, and extension. In response to local needs, the station carries out programs on the improvement of varieties, cultural practices, and the use of pesticides, fertilizers, and herbicides. These stations also cooperate with the TARI in conducting province-wide regional tests on varieties and fertilizers. The DAISs have made a number of valuable contributions to the progress of Taiwan's agriculture, including the successful breeding of new varieties of rice, corn, peanut, sorghum, millet, and onion; the cultivation of muskmelon; and techniques to produce horticultural crops under protected environments.

The extension department serves as a bridge between research organizations and the Farmers' Associations through which technical information and guidance are disseminated to the farm level in discussion meetings, demonstrations, and other special gatherings of farmers. This extension network has also encouraged researchers to work more effectively on the most significant problems for rural people. Research and extension therefore complement one another.

Conclusion

This chapter has presented the three different institutional organizations that have contributed the most to Taiwan's agricultural development. They are the Sino-American Joint Commission on Rural Reconstruction (JCRR), the farmers' organizations, and the agricultural research institutions.

The JCRR was set up as a bilateral agency on the mainland before the government moved to Taiwan. It became the major institution—comparable to a ministry of agriculture—in planning and helping to carry out Taiwan's agricultural development since 1953. The JCRR was involved in an integral way in the conduct of agricultural activities. From the outset, the JCRR was given the responsibility for planning agricultural development jointly with the Ministry of Economic Affairs and the Economic Planning Council. The planning process was clearly an indicative type; major and subsidiary objectives were targeted through a combination of broad policy instruments and specific activities and organizational changes.

One important characteristic of the JCRR that should not be neglected is that it was established as a semipublic institution to help allocate and channel U.S. aid. This feature provided it some independence from the government. The JCRR could take a longer-run view of the agricultural development process because it was insulated from the daily political and bureaucratic pressure of a typical agricultural ministry.

Agricultural development plans cannot be effectively implemented without the active support of well-organized farmers. Only when farmers are so organized can the government agricultural agencies effectively extend improved varieties of crop or livestock and render technical services, credit institutions conveniently reach individual farmers in their own operations, and collective bargaining power of small farmers be strengthened to achieve better buying and better selling. Farmers' organizations play many socio-economic roles that contribute to basic preconditions for the modernization of agriculture, especially in developing economies.

Farmers' organizations in different countries develop in different ways depending upon the social, economic, and cultural background of the particular nation. In Taiwan the major farmers' organizations are the farmers' associations, the irrigation associations, and the fruit-marketing cooperatives. They have contributed greatly to agricultural development in Taiwan by providing the following support services to farmers: (1) effective irrigation water management, (2) farm production credit, (3) agricultural extension service, and (4) efficient supply and marketing facilities.

The efforts made by the farmers' organizations to increase agricultural production and incomes under the overall economic development program have been essential prerequisites to the industrial growth of Taiwan. In fact, rural areas in Taiwan supplied most of the resources needed for industrial development during its initial stage. They also offered a huge market for industrial products. Most agro-industries cooperated closely with farmers' organizations in promoting plans for the production, processing and export of farm products.

The development of agricultural technology and a complementary set of institutions suited to underlying resource endowments has also contributed greatly to Taiwan's agricultural development. In the early stage of development—when land and capital were scarce—Taiwan relied upon an agricultural technology that maximized utilization of its abundant available labor. This technology took a package approach, using inputs that were highly correlated with labor, such as high-yielding varieties of seeds, chemical fertilizers, irrigation and improved drainage, and selective mechanization in the form of power tillers that were not competitive with labor. The very real success of this technology, in using all labor resources available, led to a serious labor shortage after 1968, which required a major switch toward mechanization and capital as a substitute for labor. Also, by pushing on different fronts, agricultural research encouraged product diversification.

The close working relationships among the JCRR, the farmers' organizations, and the agricultural research institutions efficiently linked agricultural planning, research, and extension in Taiwan. This development strategy may be found useful and potentially transferable, within limits, to other developing countries.

References

Galenson, W. 1979. *Economic Growth and Structural Change in Taiwan, the Postwar Experience of the Republic of China,* Ithaca, New York: Cornell University Press.

Hou, C.M. and Yu, T.S. (ed.). 1982. *Agricultural Development in China, Japan, and Korea,* Academia Sinica, Taipei.

Hsieh, S.C. and Lee, T.H. 1958. An Analytical Review of Agricultural Development in Taiwan—An Input-Output and Productivity Approach, *JCRR Digest Series,* No. 12.

Mao, Y.K. 1976. "The Most Effective Strategy to Promote Rural Development—With Emphasis on the Experience of the ROC," *Industry of Free China,* Vol. 46, No. 2.

______. 1984, 1985. "Agricultural Development Policy and Performance in Taiwan, R.O.C." *Industry of Free China*, November, December and January.

Shen T.H. (ed.). 1974. *Agriculture's Place in the Strategy of Development: The Taiwan Experience*, JCRR.

______. 1970. *The Sino-American Joint Commission on Rural Reconstruction: Twenty Years of Cooperation for Agricultural Development*, Ithaca, New York: Cornell University Press.

Discussion

Liu Zhicheng

Having had the pleasure of perusing Professor Yu-Kang Mao's work, titled "Institutional Organizations in Taiwan's Agricultural Development," I believe the chapter is highly worth reading for five reasons.

First of all, in his introduction, Mao briefly describes the remarkable achievements of Taiwan's economy during the past four decades. Per capita national income reached U.S. $4,630 in 1987, nearly 30 times its U.S. $159 per capita national income in 1953. High economic growth rates have been achieved despite rapid population growth and limited natural resources. For this, may I, as a compatriot from the Mainland, extend my congratulations.

Secondly, Mao briefly describes the process of Taiwan's transformation from a predominantly agricultural economy, with exports relying heavily on sugar and rice, to a fairly diversified economy, with industry the leading sector and with a tendency for change in agricultural production. It seems to me that these tendencies in agricultural production and in the overall economy of Taiwan are similar to those of the Mainland in many ways. For example, during the past several decades, from 1953 to 1987, the share of agriculture in the structure of the national economy dropped and that of industry increased year after year both on the Mainland and in Taiwan. Within the overall structure of agricultural production, the reduction of the share of crop production and the increase in the share of animal husbandry and fishery continued (see Table 5.3).

Why is the tendency of changes in both industrial and agricultural production in Taiwan generally similar to that of the Mainland? My answer is that the basic situation of industrial and agricultural production on both sides of the Strait is similar, for both economies are in the developing stage. More importantly, we share the same ancestors and origins, with similar agricultural production patterns, life styles, culture, and daily customs.

Thirdly, in his work Mao introduces in detail the history and development of three agricultural organizations (the JCRR, the farmers' organiza-

TABLE 5.3 Tendency of Changes in Agricultural Production (%)

	Taiwan		*Mainland China*	
Item	*1953*	*1987*	*1953*	*1987*
Crop production	71.9	41.4	81.8	60.7
Forestry	5.1	1.2	0.7	4.8
Animal husbandry	7.4	29.2	12.6	22.8
Fishery	15.6	28.2	0.3	4.7
Side-line occupation	-	-	4.6	7.0
Total	100.0	100.0	100.0	100.0

Source: Council of Agriculture, *Basic Agricultural Statistics*, Republic of China, 1988 and statistics from the State Statistical Bureau, Beijing.

tions, and the agricultural research institutions) and their roles in promoting agricultural production. I think the key reason why those three agricultural organizations in Taiwan became important forces in promoting agricultural development is that they combined the planning function with the actual implementation of research and extension. Because both of these activities were linked, they benefited one another. It also seems to me that agricultural organizations in Taiwan are well developed, forming a horizontal and vertical network with three administrative levels, namely, provincial, county, and township. The structure serves farm members of the organizations in technology extension, finance and credit, and transporting and marketing products, and helps them to overcome the difficulties in technology research and delivery. In fact, agricultural organizations are multipurpose cooperative organizations.

I also fully agree with Mao's point of view with respect to the functions and roles of agricultural organizations. For example, he affirms that agricultural development plans cannot be effectively implemented without the active support of the farmers who must be well organized to effectively receive improved crop varieties or better livestock breeds to benefit from technical services. Only within a context of effective farm organizations can credit institutions conveniently reach individual farmers in their own operations. The collective bargaining power of farmers can also be strengthened to achieve better bargains in both buying and selling.

Fourthly, in reading Mao's work, I feel that agricultural research institutions in Taiwan are also relatively well developed. These institutions are divided into four categories (provincial agriculture research institutions, industrial research institutions, agricultural colleges, and independent research institutions), with different divisions and priorities. Together, these research institutions have addressed the practical reality of agricultural production, studied the technical problems in production urgently needed by farmers, formed a research network throughout Taiwan, and have been an important force in promoting its development of agriculture. In addition, the District Agricultural Improvement Stations (DAIS) are vital to local farming research and extension. They serve as a bridge between research organizations and the Farmers' Associations, through which technical information and guidance are disseminated to the farmers on the one hand and difficult technical problems among farmers are passed to research organizations on the other. Thus, research and extension complement one another.

Finally, my conclusion from reading Professor Yu-Kang Mao's work is that agriculture in Taiwan developed rapidly with abundant labor but with scarce land resources. All three groups of agricultural organizations have played a very significant role in Taiwan's overall development; although, of course, agricultural development is supported by many other factors. Taiwan's precious experiences in promoting agricultural development through integrated agricultural organizations are well worth being studied and used as a reference.

Against this backdrop, I would like to very briefly introduce agricultural research institutions and extension organizations in the Mainland.

1. Three levels of agricultural research institutions.
   a. National agricultural institutions such as the Chinese Academy of Agricultural Sciences. This is the national center of agricultural research with applied fundamental research as the main task. The institution organizes and collaborates research programs and service activities, and also provides academic guidance throughout the country.
   b. Provincial agricultural research institutions. These institutions, with applied research and development as the main tasks, guide and collaborate research throughout the province. The institutions also partly undertake national and regional research responsibilities.
   c. Agricultural research institutions at the levels of the prefecture and the municipalities. These institutions mainly undertake developmental research, as well as select priority research programs where conditions permit.

   In addition, dozens of agricultural universities and colleges, several agricultural enterprises (such as the reclamation system), and a number of research institutions in the Chinese Academy of Agricultural Sciences also engage in considerable research on agricultural problems.

2. Four levels of agricultural technical extension organizations.
   a. National agricultural technical extension organizations under the Ministry of Agriculture. These include general research in the areas of stations of seeds, plant protection, soil and fertilizers, animal husbandry and veterinary science, agricultural technical extension, and business management.
   b. Provincial agricultural technical extension organizations. They are generally the counterparts of national extension organizations.
   c. County agricultural technical extension organizations. Special technical extension organizations in almost all counties have been reorganized into an extension center of agricultural technology. These centers are responsible for agricultural and animal husbandry experiments, demonstration, training, and extension of agriculture and animal husbandry.
   d. Extension levels of township and village. Comprehensive or agricultural special extension organizations such as technology extension stations (or groups), have been set up where necessary.

According to the statistics resulting from a science and technology survey conducted in 1985 on the Mainland, there are 1,169 agricultural research institutions with a total staff of 133,000, among whom 95,000 are researchers. There are 119,700 extension organizations for agricultural technology at various levels, with a total staff of 679,000, among whom 247,200 are technicians. Another 80,000 agricultural organizations of different kinds are run by local people with a total technical staff of 280,000. In addition, 500,000 farmer technicians operate at the township and village levels.

This brief introduction demonstrates that, on the Mainland, a fairly complete system has been formed to link agricultural research and extension organizations from the top to the grass roots levels. The system promotes increased agricultural production and the adoption of new technology. It also completes the efficient chain of fundamental and applied research, developmental research, and technology extension.

PART THREE

Prices, Costs, and Incomes

Newly harvested corn at a farm in China. *(Courtesy Frederick Crook)*

Introduction to Part Three

Peter Calkins

Within the general institutional framework of the preceding section, we now turn to an evaluation of three specific questions: What is the impact of macro and agricultural policies upon relative input prices? How does the choice of technology and institutional structure affect production costs? What is the role of both policies and structure in determining the levels and distribution of net income of farmers? Because we will often wish to answer these questions in terms of either specific markets or individual households or production units, the analysis will be conducted at both the macro- and the micro-economic levels.

Prices

Chapter 6 by Tso-Kwei Peng responds to the first question by outlining the impact of Taiwan's agricultural price policies, notably the terms of trade for agriculture, designed to extract a steady agricultural surplus to foster the ultimate goal of industrial development. Since 1973, these terms of trade have been switched to favor agriculture. During the unfavorable period of 1949-73, the agricultural sector was squeezed through such measures as the rice-for-fertilizer barter scheme and a rationing program to the armed forces; interestingly, these policies bore some similarities to the mechanisms used on the mainland. But according to Peng, what counted on Taiwan was the exact fixing of relative prices on operational markets with a separate price regime for each product.

In his comment, however, Guangsen Wang challenges the implication that turning the terms of trade in favor of agriculture is a good policy for reducing rural-urban income inequalities and promoting the rational inter-sectoral allocation of incomes. He also clarifies that population pressure and unemployment in Taiwan in the early 1950s made the supply of cheap food even more critical to social security and political stability. He notes that these policies worked not only because of the general LDC characteristic of a relatively isolated agricultural sector, but also because of Taiwan's unique economic, financial, technological, and institutional factors. Thus, the Taiwan experience may not necessarily be easily transferable.

Against these very pertinent nuances, Peng points to the success of these policies that led to both an 8.8 percent annual economic growth between 1953 and 1988 on Taiwan and positive impacts on dietary quality and diversity. At the same time, the inability of even a modernizing agricultural production system to keep up with aggregate food demand has resulted in higher farm prices and thus lower subsidies and policy costs than in other middle-income economies. Peng believes these facts indicate that Taiwan's food and agricultural policies will be socially effective in the long run.

In apparent contrast to these general benefits and occasional costs associated with a market orientation on Taiwan, Chapter 7 by Xi-Ji An argues that the mainland provinces' rural economic system is a mixture of markets, planning, economic controls, and self-sufficiency. Although the control of land has passed to the household, there is considerable tenurial and thus investment uncertainty surrounding 15-year land contracts. Mainland farm size increases and crop output decisions are, moreover, restricted with respect to their counterpart production units in Taiwan. Although markets have gained a role in increasing numbers of commodities, truly free markets do not exist because of administrative interference.

However, Carson Kung-Hsien Wu replies that, to some degree, a mixed economy has been revealed on both sides of the Taiwan Strait. Indeed, in the early stages of growth, Taiwan adopted taxation and price distortions to transfer agricultural surpluses to other sectors of the economy. Notably, production and marketing of rice, the principal food grain and foreign exchange earners in 1950-64, were tightly controlled. Moreover, Wu argues that Chinese farmers everywhere apply passive resistance to misleading and inefficient policies imposed by totalitarian methods. This explains why growth has been strong only during Type II periods of the mainland's pendulum swing, when economies of scale and economic incentives were allowed (see Chapter 2).

Within this uniquely mixed structure on the mainland, An explains that market mechanisms have witnessed the transfer of greater responsibility to new marketing agents and the emergence of multichannel marketing with only a semi-monopoly by state commerce. The problem is that the new agents are in part independent businessmen and in part administrative planners: a situation of potential conflict of interest. Their semi-interested choices have seriously affected not only agricultural prices but also production costs and farm income.

Costs

Peng argues that Taiwan has followed a pattern of substituting capital for labor, in accordance with the theory of Hayami and Ruttan of induced technical change. Because of Taiwan's market-based rural economy, both input use levels and the type of technological change have responded to changes in relative input prices. Using an aggregate production function, Peng finds that technical progress was gradually replaced by input increases as the source of growth. Nevertheless, investment levels and appropriate farm scale for mechanical innovation have lagged behind other changes because of both inflexibilities in land tenure and traditional values that cherish land as a repository of wealth. Combined with a rapid outflow of labor from agriculture, small farm size has meant that labor productivity gains have trailed far behind climbs in the wage rate.

But Wang suggests an alternative interpretation of the same statistical findings, noting that conventional inputs accounted for 60 percent of Taiwan's output growth during 1952-81; that the input structure varied between the different stages of development; that the potential of traditional technologies had been completely exhausted by 1977-81; and that in the last stage, the labor input has become totally unresponsive to increasing growth pressures on agriculture.

Peng continues his reasoning by noting that, given Taiwan's more open position on world markets than, say, Japan's, recent emphasis has been placed on farm size expansion and rural industrialization through such measures as joint corporations, cooperative farming, joint management, and the farm mechanization fund. Paradoxically, as the mainland is moving towards a Taiwan-style marketing organization, Taiwan is moving towards a mainland-style collectivization, albeit voluntary, of the production process. As a result, both the multiple cropping index and the diversification index have dropped, despite Taiwan's great multicropping potential. Indeed, Wang points out that with the increasing capital-intensity of modern production techniques, the notion of the multiple cropping index has lost much of its original meaning.

For the mainland provinces, An believes that agricultural procurement acts as a product tax in kind, of which the amount varies as procurement volume and price, market price, and total output of the specific product. After careful reasoning, he concludes that the burden of this tax falls on the farmer. For if government reduced procurement quotas or raised prices, the supply curve for agriculture would shift outward and farmers' net cash income would increase (as it did historically in 1979-84). In the meantime, farmers have been coping with heavy product taxes by switching resources from grain and cotton production to nonagricultural trade and/or to agricultural products that are taxed less heavily. If the new semi-private marketing agents reduced input prices, they would stimulate not only greater marketed supply of farm products, but also increases in net farm income.

An further believes that sustained slow growth is feasible on the mainland, but that plan evasion and administratively induced price and product tax swings make it difficult for the system to work smoothly. He calls for further reform to curb the self-interested behavior of the semi-private marketing agents.

Incomes

In the case of Taiwan, Peng contends that both income improvements and capital sources for technical change in agriculture have depended largely upon off-farm employment. Between 1970-80, off-farm income explained 89 percent of the annual 7.2 percent growth in farm family income, and grew to more than half of income from all sources. As to the distribution of this income, it demonstrated a pattern *opposite* to that of Kuznet's inverted U: by the mid-1970s, the Gini coefficient for rural income in Taiwan dipped to .28 but since has risen to .33. Peng concludes that the role of off-farm rural employment as an equilibrating force appears to be diminishing as the economy evolves toward larger farm sizes and increased mechanization. What will be needed to reduce rural income equalities in the future is programs to encourage out-migration from farming as large-scale mechanization proceeds.

While recognizing the role of off-farm income in Taiwan, Wang stresses the importance of differentiating Taiwan's part-time farm households into two categories, those that receive more than half, and those that receive less than half of their incomes from farming. Only the former are willing to directly invest resources in upgrading land productivity. Wang believes, contrary to Peng, that increased off-farm earnings should **reduce**, rather than increase income disparities. Wang argues that such measures as land consolidation, joint operations, entrusted farming, and cooperative farming may improve farm efficiency but can in no way allow farm families to achieve nonfarm family income levels. He also warns against the serious drawbacks of a collective system of land ownership, suggesting that Peng's advocacy of such a position is too facile.

For the mainland provinces, An points to the income distributional impacts of two types of policies. On the production side, he argues that the failure to take into account differing levels of managerial ability in the initial allocation of farmland (differences that would normally be expected to lead to unequal incomes across households), combined with the lack of land-right lead to transfer mechanisms, has led less to income equity than to inefficient resource use. Extremely small holding size (0.6 ha) and tenurial uncertainty have exacerbated the diseconomies of scale that heighten inefficiency. Although income is nominally better distributed, both the farmer and society lose from these policies.

On the marketing side, plan evasion by farmers distorts the income distribution targets of the national plan. In contrast, plan evasion by the semi-private commercial agents distorts incomes largely in their own favor.

More generally, Wu points to the common need for improvement in the rural development structures of both Taiwan and mainland China, notably in the areas of unrealized economies of size, sluggish out-transfers of labor from agriculture, under-used cooperatives, and persistent pollution upstream and downstream from the production unit. He therefore suggests the great benefits to be gained from continued exchange of rural development experiences between the two parts of China. For example, the recent experience of major structural change on the mainland has led to greater cropping diversity, which has had a positive ecological impact. Taiwan could benefit from deeper study of this experience.

In the other direction across the Taiwan Strait, Wu argues that Taiwan's experiences can be applied for further improvement of agricultural development on mainland China in three ways. First, a well-established education system such as Taiwan's could promote the fuller utilization and higher quality of rural labor. The benefits of such a system were maximized in Taiwan by a cost-sharing program between government and farmers for rural public works and community development projects. Second, rural industrialization, rural urbanization, and the increased use of motorcycles seem to have been appropriate and effective dynamic forces for Chinese rural development. Finally, the Taiwan case seems to demonstrate that professional development experts and graduates of vocational schools well grounded in advanced technology and theories are the best agents in society to provide expert assistance.

6

Prices, Income, and Farm Policy in Taiwan

Tso-Kwei Peng

Introduction

The objectives of this chapter are to analyze the impacts of Taiwan's agricultural policies upon the relative level of agricultural prices and hence on the allocation of resources; to assess the effects of relative price levels for traditional and nontraditional factors upon incentives for technical change and input substitution; and to investigate the sources and distribution of farm family income throughout the process of economic development.

The key strategy underlying Taiwan's economic policy has always been to "foster industrial development with agricultural resources and further agricultural prosperity with industrial sinews" (Council of Agriculture, 1986). Policies to change the terms of trade for agricultural products were used as a main part of this strategy during the process of economic development. Agriculture was the mainstay of Taiwan's economy in the 1950s through the early 1960s. The rapid growth of agricultural production through institutional and biological innovations not only produced sufficient food for the nation but also generated surplus capital and released productive labor for development projects in non-agricultural sectors. Contributing substantially to the nation's efforts at speeding up industrialization and promoting overall economic progress, agriculture has indeed successfully performed its function of fostering industrial development.

With the rapid structural changes in the economy, the value of industrial production exceeded that of agriculture by the mid-1960s. Several structural adjustment issues emerged, resulting in sizable increases in agricultural production costs. The competitiveness of farm products thus declined, the rate of agricultural growth fell, and farm family incomes did not keep pace with urban incomes. In light of these agricultural problems arising from changes in economic structure, the government has switched its policy from "development of agriculture to foster industry (taxing producers)" to "development of industry to support agriculture (subsidizing farmers)." Since 1973, a series of farm development programs were put into effect to revitalize agriculture. The present chapter includes a review of the changes in terms of trade for agricultural products as well as agricultural price and income policies; an analysis of the characteristics of input-output relationships in Taiwan's agriculture; and an evaluation of the sources and distribution of farm family income in Taiwan.

Changes in the Relative Price Level of Farm Products in Taiwan

To achieve stable and sustained economic growth, a proper balance between agricultural and industrial sector growth rates should be maintained. Since World War II, agricultural terms of trade have improved in Taiwan, if measured by the long-run trend. The data are shown in Table 6.1 and Figure 6.1, where each year's agricultural price index is alternatively divided by the industrial and wholesale price indices for the base year 1971.

The prices of farm products increased relative to prices of nonfarm products during 1945-1949, 1973-1976, and again after 1981. In general, the terms of trade were unfavorable to agricultural products before the early 1970s, but this phenomenon changed after 1973. In the early stage of economic development in Taiwan, the government adopted price distortion and taxation approaches to siphon agricultural surpluses to nonagricultural sectors. Traditionally, rice is a staple and the most important food grain; therefore, the production and marketing of rice were always tightly controlled. The government implemented the rice-for-fertilizer barter system with an exchange ratio unfavorable to rice farmers, compulsory government procurement of paddy rice at a lower-than-market price, and a paid-in-kind land tax to skim off the agricultural surplus. Rice collected was rationed to the armed forces and distributed to military dependents and civilian government employees. Moreover, the stocks were either sold on the free market to stabilize rice prices or exported. The terms of trade, therefore, were unfavorable to agricultural products, particularly in the case of rice, from 1938 to 1972, except during the hyperinflationary period of the early 1940s.

Since 1973, the government has implemented a series of rural development programs to speed up the process of agricultural modernization. The general framework of current agricultural policy in Taiwan can be summarized in Table 6.2. Instead of implementing a lower price policy for rice, the government has enacted price support and stabilization schemes. The price

TABLE 6.1 The Relative Price of Agricultural Products in Taiwan (Base Year: 1971)

Period	*Agricultural Product* %	*Rice* %
1938-44	67	74
1945-49	108	89
1950-60	76	70
1961-72	93	94
1973-76	110	119
1977-81	97	114
1982-84	100	106
1985-88	110	112

Source: Directorate-General of Budget, Accounting & Statistics, Executive Yuan, *Monthly Report of Price Statistics, Taiwan District*, Taipei, Taiwan, 1989.

FIGURE 6.1 The Trend of Relative Prices for Taiwan's Agriculture Products

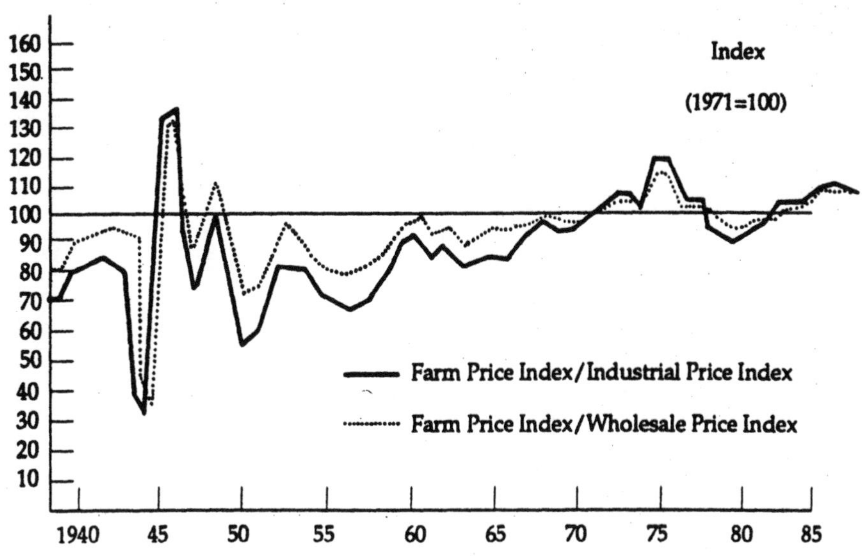

TABLE 6.2 Overall Framework of Taiwan's Agricultural Policies

Policy Area	*Aims*	*Major Projects/Measures*
Structural policy	Expansion of farm size Encouragement of joint farming	Second stage of agricultural land reform program Farmland purchase scheme Regional planning program Nucleus farmer program Encouragement of rural industrialization
Production policy	Adjustment of agricultural production structure Promotion of agricultural efficiency and self-sufficiency on major production Improvement of agricultural infrastructure	Paddy field conversion program Research and extension Water and land resource development Farm mechanization scheme Rural public investment
Price and income policy	Price stabilization and support Farm family income enhancement	Price stabilization scheme for rice, small grain, sugar, major crops Livestock insurance Agricultural development scheme and loan program Border measures
Marketing policy	Rationalization of marketing channels Improvement of food quality	Agricultural marketing act Food processing development program Joint marketing program Wholesale market modernization program
Rural life policy	Improvement of environmental factors in rural villages Improvement of farmers' welfare	Farmer health insurance program Rural village reorganization scheme Establishment of farmer shopping centers

of farm products, especially rice, has increased relative to the prices of nonfarm products since 1974. In 1977-1981, agricultural prices dropped sharply because of both higher livestock marketing in the business cycle throughout the period and the influx of foreign agricultural products.

The relative growth rates of supply and demand are another factor affecting the movement of the relative price for farm products. Generally speaking, the growth in demand for any commodity or service depends upon the growth rates of the population and of real per capita income, and the income elasticity of demand for any particular commodity or group of commodities (Johnson, 1973). An increase in per capita income means not only that a smaller percentage of income is spent on food (Engel's Law), but also that the pattern of food consumption changes. According to what is usually known as Bennett's Law (Timmer, Falcon, and Pearson, 1983), the proportion of calories an individual consumes from the basic starchy staple is expected to decline. This tendency has a considerable impact upon agricultural production and its structure.

In Taiwan, the annual economic growth rate was 8.8 percent from 1953 to 1987. The proportion of food expenditure to household consumption expenditure decreased from 54.4 percent in 1953 to 30.4 percent in 1987. The consumption of starchy foods has dropped considerably over the last two decades in Taiwan, while the indirect consumption of grain embodied in high quality foods has increased dramatically (Figure 6.2). According to the "Food Balance Sheet" published by the Council of Agriculture, per capita annual consumption of rice and sweet potato fell significantly, while annual per capita consumption of meat, eggs, dairy products, vegetables and fruit rose significantly from 1953 to 1987. As for the per capita consumption of grain (including rice, wheat, corn, soybean, barley, and sorghum), direct consumption increased before 1974 and then dropped. Indirect per capita consumption of grain, on the other hand, rose consistently, causing total grain consumption to increase from 171.0 kg in 1953 to 411.4 kg in 1987. But calories from starchy staples have declined (see Figure 6.3).

Both the population growth rate and the income elasticity of food demand declined between 1962 and 1981 in Taiwan. The growth rate of aggregate demand for food, therefore, also slowed. However, rapid changes in food consumption patterns, and natural constraints on producing high quality food imply that agricultural production has grown less rapidly than either the growth rate of aggregate demand for food (Table 6.3) or output in the rest of the economy. As a result, farm product prices increased rapidly relative to nonfarm product prices, and substantial quantities of agricultural products have been imported.

As in many other developed economies, agricultural price and income policies play an important role in Taiwan. The objectives of price and income policies are to stabilize and support agricultural prices at reasonable

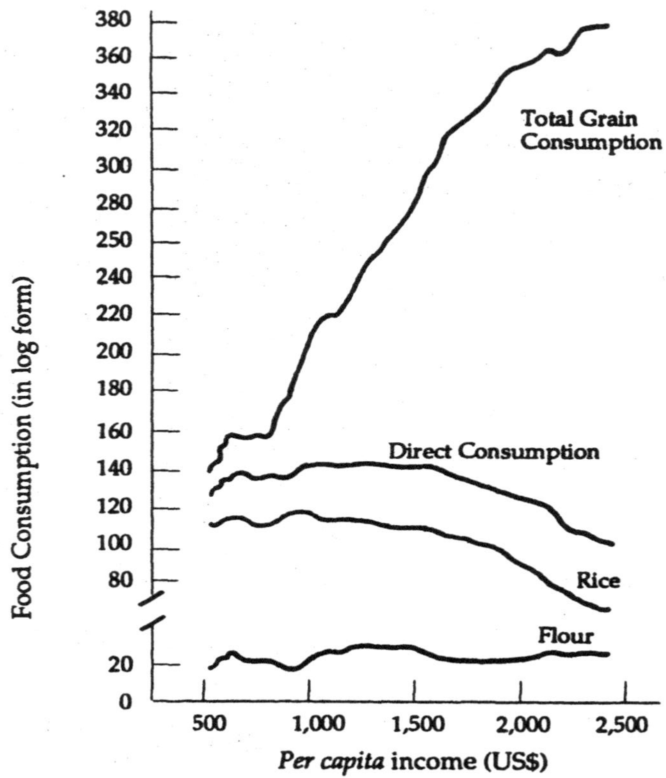

FIGURE 6.2 Per Capita Consumption of Grain Relative to Income, Taiwan

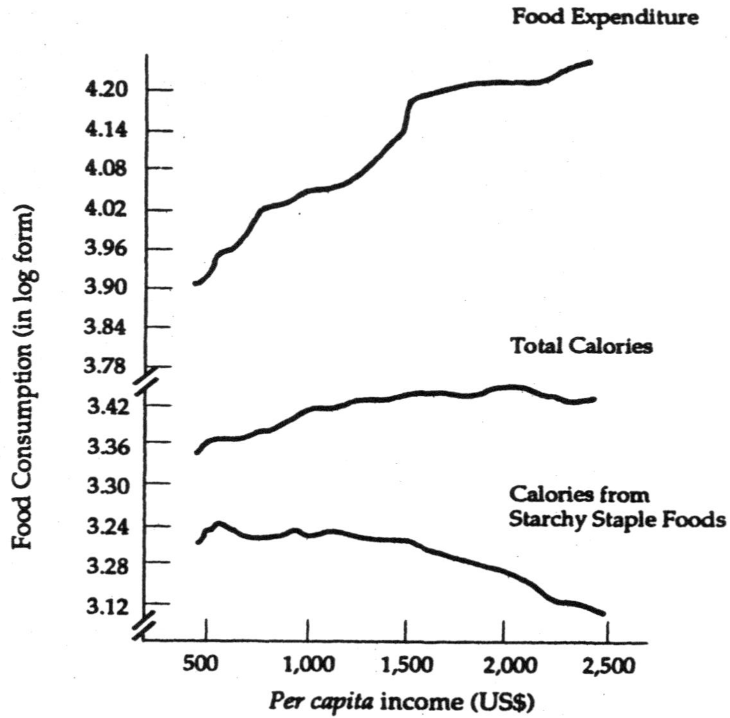

FIGURE 6.3 Food Consumption in Relation to Income Level, Taiwan

TABLE 6.3 Growth Rate of Food Demand and Agricultural Production

	Average Growth Rate			*Average Growth Rate*	
Period	*Population*	*Income*	*Income Elasticity*	*Demand*	*Production*
1962-66	3.1	5.4	0.5	5.8	5.8
1967-71	2.9	7.8	0.4	6.3	5.7
1972-76	1.9	5.7	0.4	4.1	3.3
1977-81	1.9	4.9	0.3	3.6	1.1

Source: Peng, T. K. (1988).

levels, and to reduce production costs and the financial burden associated with subsidies. The major policy instruments used are price support programs, price stabilization programs, direct payments, and input subsidies. Taiwan's price and income policies after 1973 can be summarized as follows:

1. Each product or product group has its own specific price regime, reflecting the nature of the product and its historical background. However, only the rice-price-support and the rice-field-conversion programs are financially supported by the government. The price schemes for fruits, vegetables, and livestock products are financially self-sufficient. These schemes, mainly operated by farmer's associations and cooperatives, have a very weak financial base. Because the free-rider problem of collective action is uncontrollable, the bargaining power of these organizations appears weak, and their price schemes ineffective.
2. The level of agricultural subsidization in Taiwan is much lower than in other developed economies. Taiwan's agricultural budget in recent years accounts for 2-3 percent of total government expenditures (Chen, 1987). The budget allocates funds for infrastructural improvements and disaster restorations, agricultural research and extension, and price supports for rice and paddy field conversion programs. The producer's subsidy equivalent (PSE) of Taiwanese major land-based crops, from 1982 to 1986, ranged from 30 to 70 percent. The average PSE in Taiwan was 19.2 percent, much lower than Japan's 71.1 percent, Korea's 59.5 percent, the EEC's 35.4 percent, Canada's 31.0 percent, New Zealand's 25.4 percent and the U.S.A.'s 24.6 percent (USDA, 1988).

3. The social cost of government policy in Taiwan is significantly less than that in other developed countries. Accordingly, policy costs, in terms of the dollars of government and consumer payments required to increase producer's surplus by $1, are also lower in Taiwan. For example, the policy cost for rice in Taiwan was $1.29 in 1986 (Peng, 1988), lower than Japan's 2.58 in 1976, the EEC's 1.50 in 1980 and the USA's 1.38 in 1985 (World Bank, 1986). The government and consumer shares for the cost increase in producer's surplus are 56 percent and 44 percent, respectively.

Technological Change and Input Substitution in Taiwanese Agriculture

Taiwan's rapid economic development since the mid-1960s has changed the structure of both agriculture and farm expenditures. Cash expenditures, which accounted for about 66 percent of total farm expenditure in 1960, rose to 90 percent by 1987. This jump signals Taiwan's transition to a modern commercial agriculture.

The economic growth of Taiwan has changed the relative price of human inputs to such nonhuman inputs as fertilizer, pesticides, and farm machinery. The agricultural growth in Taiwan has originated from improvements in the quality of human and nonhuman resources, with capital substituting for labor (Chen and Wang, 1980). The sustained industrialization and urbanization that occurred in Taiwan over the last two decades have created abundant nonfarm employment opportunities for farm labor. Wages in agriculture increased substantially, relative to the prices of nonhuman inputs during the 1970s, a period of unusually rapid growth in nonfarm employment. The annual rate of price change for different inputs at various times was in accordance with both general and farm product price levels. Specifically, rapid increases occurred from 1951 to 1960 and again from 1973 to 1981. Since 1982, input prices have registered a decreasing trend, while wage rates rose 3.4 percent annually through 1987.

As a result, the price of nonlabor inputs relative to the farm wage rate in 1987 was only one-tenth of that in 1952. Relatively cheaper capital inputs, therefore, were substituted for farmland and farm labor. Consistent with these changes in farm wages, labor input per farm decreased 3.3 percent annually from 1960 to 1987. At the same time, current capital inputs—including fertilizer, agricultural chemicals, feed, and water—increased about 3.8 percent due to technological change, relatively cheaper current capital inputs, and the expansion of livestock production. Farmland crop area shows a decreasing trend since 1970, mainly due to more extensive production techniques. Fixed capital inputs (including farm buildings, equipment, perennial crops, and animals) rose 2.6 percent annually, more slowly than current capital inputs. General farm expenditures rose 8.3 percent annually from 1960 to 1987.

Both input use and technological change responded to the changes in relative input prices. These adjustments have led Taiwan's agricultural production to depend increasingly on modern inputs and foreign markets. To evaluate the structure of aggregate agricultural production and the characteristics of input substitution in Taiwan, we have analyzed (Peng and Lin, 1989) aggregate output and input data from 1952 to 1982 (Chen, 1984) using the translog cost function (Christensen, Jorgenson, and Lau, 1973). The translog cost function places no *a priori* restrictions on substitution possibilities among the factors of production, and allows economies of scale to vary with the level of output. This function is specified as:

$$\ln C = \gamma_o + \gamma_Y \ln Y + \gamma_T T + \gamma_{YT} T \ln Y + \Sigma\gamma_i \ln P_i + \Sigma\gamma_{iY} \ln Y \ln P_i$$
$$+ \Sigma\gamma_{iT} T \ln P_i + \Sigma\, \Sigma\gamma_{ij} \ln P_i \ln P_j + 1/2\gamma_{YY} (\ln Y)^2$$
$$+ 1/2\gamma_{TT} (T)^2$$

where $\gamma_{ij} = \gamma_{ji}$, C is total cost, Y is output, the P_i's are the prices of factor input, and T is time trend. For this function to be well-behaved, it must be homogeneous of degree one in prices.

Using Shepherd's lemma, the input demand functions for the factors of production can be derived by partially differentiating the translog cost function with respect to factor prices, that is,

$$\partial \ln C / \partial \ln P_i = P_i X_i / C = S_i$$

where S_i denotes the cost share of the ith-factor input. The translog cost function yields the following cost-share equations:

$$S_i = \gamma_i + \sum_{j=1}^{N} \gamma_{ij} \; P_j + \gamma_{iT} \; T + \gamma_{iY} \ln Y$$

Allen's partial elasticities of substitution can be computed from the translog cost function as:

$$\sigma_{ij} = (\gamma_{ij} + S_i S_j) \;/\; S_i S_j \qquad \text{for } i \neq j$$

$$\sigma_{ii} = (\gamma_{ij} + S_i (S_i - 1)) \;/\; S_i^2 \quad \text{for } i = j$$

The impacts of technological change upon cost structure can be computed by partially differentiating the translog cost function with respect to the trend variable (T), that is

$$\partial \ln C/\partial T = \gamma_T + \gamma_{TT} T + \Sigma\gamma_{iT} \ln P_i$$

where γ_T and γ_{TT} T constitute the Hicks neutral technological change component, and $\Sigma\gamma_{iT} \ln P_i$ represents the biased technological change component.

We specified additive disturbances for each share equation and for the cost function. The results of using the iterative, seemingly unrelated approach to estimate the coefficients were quite satisfactory (Peng and Lin, 1989). The aggregate production function was also estimated to assess the sources of agricultural output increases.

In general, agricultural production can be continually expanded to respond to changes in relative input prices. The growth of agricultural output derives mainly from technological changes and increases in input use. Technological innovations have been a vital factor in Taiwan's agricultural development. From 1952 to 1982, agricultural technical progress was mainly attributable to neutral technical change, but the importance of the latter declined in the process of agricultural growth. In the early stage of development, from 1952 to 1966, induced innovation from changes in relative input prices caused aggregate production costs to increase because agricultural input prices were biased against technological change. However, biased technical change became an important factor in increased agricultural output from 1967 to 1982. The contribution of technical change to the expansion of agricultural output in Taiwan declined during the period of study. The share of agricultural technical change in total agricultural output growth decreased from 46.6 percent to 32.8 percent (Table 6.4). In other words, the annual increase in agricultural output was mainly due to a drastic increase in input utilization, especially in the later stage of agricultural development.

The data in Table 6.4 reveal several facts worth mentioning. First, the major factors contributing to Taiwan's agricultural growth were dramatic increases in current input utilization and technological change during the study period. Second, the growth rate of agricultural output declined after 1967 mainly because of decreased labor and land inputs and relatively slower growth in technological change than in the earlier stages. However, current inputs were intensively used to keep the livestock sector growing. Third, fixed capital inputs made only a small contribution to the growth of agricultural output. These facts demonstrate that investment levels and appropriate farm scales for mechanical innovation still lagged far behind in Taiwan. Finally, the trends in prices paid and received by farmers were

TABLE 6.4 The Annual Growth Rate of Agricultural Output

	Unit: %		
Items	*1952-1967*	*1967-1981*	*1952-1981*
Agricultural output	4.5 (100.0)[a]	3.4 (100.0)	3.9 (100.0)
Technical change	1.9 (46.4)	1.3 (32.8)	1.6 (43.7)
Neutral	2.0	1.0	1.5
Biased	-0.1	0.3	0.1
Labor input	0.5 (10.1)	-1.4 (-40.9)	-0.5 (-12.2)
Current capital	1.7 (36.8)	3.5 (105.2)	2.5 (64.5)
Fixed capital	0.1 (2.6)	0.3 (7.8)	0.2 (5.1)
Land input	0.2 (4.2)	-0.2 (-4.8)	-0.04 (-1.0)

[a]The number in ( ) is the contribution rate of a specific factor to the growth of agricultural output.

Source: Peng and Lin (1989).

similar during the period of study. However, the contribution of current capital inputs to the growth of agricultural output increased from 36.8 percent in the early period to 105.2 percent later. This implies that the growth rate of agricultural value added was smaller than that of agricultural output. Thus, the functional income distribution squeezed traditional inputs, such as farm labor and agricultural land.

As far as input substitution is concerned, substitution occurred between labor and current capital inputs, and current capital inputs and fixed capital inputs, as shown in Table 6.5. The farm labor force decreased drastically during the last two decades, and the significant increase in the farm wage rate on agricultural production was partly offset by the industrial sector's relatively cheap modern inputs. These trends induced structural changes in agricultural production. Therefore, during the modernization of farm operation, the effect of current capital inputs was greater than that of fixed capital inputs, and the growth rate of fixed capital inputs was much lower than that of current capital inputs. Growth in farm size was impeded by the highly complementary relationship between mechanical innovation and farm size, as well as an inflexible regulation of land tenure following

TABLE 6.5 Allen's Partial Elasticities of Substitution of Input Demand[a]

Elasticity[b]	*1952-1966*	*1967-1981*	*1952-1981*
σ_{LL}	-0.8975 (0.0958)*	-0.8511 (0.0935)*	-0.8609 (0.0947)*
σ_{LC}	1.2395 (0.2405)*	1.1585 (0.1591)*	1.1910 (0.1917)*
σ_{LK}	0.0640 (0.4090)	-0.1158 (0.4876)	-0.0181 (0.4449)
σ_{LA}	0.5437 (0.3042)	0.1239 (0.5841)	0.4010 (0.3994)
σ_{CC}	-3.3459 (0.7809)*	-1.9087 (0.3505)*	-2.4843 (0.5026)*
σ_{CK}	3.1477 (1.4104)*	2.7367 (1.1404)*	2.8857 (1.2384)*
σ_{CA}	-0.3012 (0.8483)	-0.6946 (1.1047)	-0.3790 (0.8990)
σ_{KK}	-9.0500 (3.1624)*	-10.5593 (4.6060)*	-9.7634 (3.7881)*
σ_{KA}	-0.1081 (1.4869)	-1.5995 (3.4881)	-0.6021 (2.1498)
σ_{AA}	-0.8122 (1.2130)	2.5649 (4.5834)	-0.2180 (2.1168)

[a]The number in ( ) is standard deviation, and * indicates the coefficient is significant at the 5% level.

[b]L = Fixed Labor; K = Capital; C = Current Capital; and A = Land

Source: Peng and Lin (1989).

postwar land reform in Taiwan, and the long-standing tradition of keeping land as an asset in rural society. Allen's partial elasticity of substitution between land and fixed capital input was not significant. Therefore, given the emergence of economies of scale, both institutional rigidity and cultural constraints slowed the ability of mechanical innovations to absorb increased production costs arising from the sharp decrease in farm labor since the 1970s. All of these have turned Taiwan's agricultural trade balance from a surplus to a deficit since 1970.

The unavoidable productivity lag due to economic structural changes after rapid agricultural development causes much more difficult adjustment problems for countries with small farms than countries with large farms (Peng and Chang, 1983). A result of the accelerated outflow of agricultural labor during the 1970s in Taiwan was an 18.7 percent annual wage increase, doubling the 8.8 percent rate of the 1960s. This was the major factor contributing to increases in crop production costs in the 1970s. Small farms are plagued by lagging farm mechanization, particularly for dryland crops. Individual farms of 1 ha are too small economically to absorb investments in small machines.

Labor productivity in the 1970s increased significantly because of the drastic farm labor outflow combined with increased modern input use. However, labor productivity still trailed far behind increases in the wage rate. Such a severe lag in labor productivity, with the same root causes as the sharp rise in farm wages, would not have happened in the following instances:

a. In countries with large farms where the impact of decreasing farm labor is offset by farm consolidation and mechanization. The existence of productivity lags on small farms make their products uncompetitive with developed countries such as the U.S., which have already experienced an outflow of farm labor, have stabilized farm wages at a high level, and have kept productivity increases ahead of rising production costs.
b. In less developed countries, which have not yet experienced a rapid outflow of farm labor, have maintained crop yields at relatively low levels, and have kept the increases in farm wages below the increases in productivity. In terms of a revealed comparative advantage index, Taiwan's agricultural products had a smaller comparative advantage than either developing countries or developed countries as a whole (Anderson, 1983).

In countries with small farms, rapid, multiphased structural adjustment makes agricultural development and resource utilization extremely vulnerable. Japan adopted a high protection policy during the 1960s and the 1970s to ensure that farmers would be protected by consumers. Instead, Taiwan reduced many tariff and nontariff trade barriers. The increasing trade surplus, particularly with the United States, prompted further efforts to liberalize trade, including that of agricultural commodities. As a result, agricultural imports increased for final consumption goods. Agricultural production came under pressure to compete both for domestic industrial sector resources, and for foreign agricultural imports. Given the unavoidable productivity lag of the agricultural sector in periods of rapid economic structural adjustment, farm operating scale was expanded to increase agricultural production efficiency, and new industries were encouraged in rural areas to raise farm family incomes.

Farm Family Income and Its Distribution in Taiwan

During the industrialization process, farmers' income increases as productivity increases. Because agricultural production is a biological process and agricultural products generally display both inelastic supply and demand, increases in farm income are usually less than those of nonfarm income, especially for small farms. A low ratio of farm to nonfarm income suggests the possibility of off-farm employment to improve income. Farm family income then comes from both on-farm and off-farm sources; thus, the potential for improving farm family income is intricately interwoven with opportunities in both agriculture and off-farm employment. In this section, the present agricultural policies stimulating Taiwan's off-farm income will be discussed. The growth and distribution of farm family income will also be analyzed to reflect the overall performance of Taiwan's farm economy.

Programs Stimulating Off-Farm Income

The basic structural problem affecting Taiwan's agriculture is the extremely small size of farms. Such size is insufficient to maintain the family and gain maximum economic efficiency. As laid down in the Agricultural Development Act, the expansion of farm size is a prime policy objective. The Second Stage of the Agricultural Land Reform Program, officially proclaimed in April 1980, encouraged farm size expansion to 15-20 ha through joint operations, entrusted farming, cooperative farming, and joint management. Such organized group farming has at least six advantages. First, it enlarges the land base without affecting land ownership or losing mechanical operating efficiency. Second, by maintaining a high level of land productivity, group farming results in lower production costs per unit of total output. Third, it helps increase farm family income by diverting surplus farm labor toward better-paying jobs in the nonfarm sector. Next, it avoids the landlord-tenant relationship by having the land entrusted to other members of the group for cultivation. Fifth, it enables small farms to perform with large-farm efficiency, even in buying and selling. Finally, it maintains the family farming system as a basis for small farm development (Ong, 1984).

The program also set up the Farm Mechanization Fund to accelerate machinery adoption, and the Farmland Purchase Fund to encourage individual farmers to expand by buying land from neighbors. Land consolidation is a major factor in the program to improve the transportation, irrigation, and drainage facilities essential to technological innovation. The second stage of the land reform program attempted to raise productivity within the context of rapid industrialization (Li, 1988), while the first stage (1940s) was mainly designed to equalize land ownership distribution.

The development of transportation, communication networks, and public utilities in the 1950s provided favorable conditions for rural industrialization. Under the Accelerated Rural Development Program (ARDP), labor-intensive industries—primarily food processing and textiles—were encouraged to establish new plants in rural areas. The government established "Industrial Zones," for which infrastructure was provided. By 1976, the total number of plants had reached 431,000, with a spatial distribution of 36 percent urban and 64 percent rural (Peng, 1987).

The absorption of workers from farm families by rural industry greatly altered the pattern of labor and land use within farm families and rural communities. Group farming systems operate highly mechanized farms, releasing farm laborers to commute regularly to factories and spend the rest of their time (peak farming seasons and weekends) working on the farm. On the average, in 1986, farmers consecrated only 30 percent of their available time to agricultural production (Peng and Chen, 1986). Rural industrialization provided rural labor with opportunities for earning attractive income in the nonfarm sector, thus greatly increasing farm family incomes.

However, the intensity of land use declined and farms became more specialized and market-oriented. This phenomenon can be examined in Table 6.6. The multiple cropping index declined from a peak of 188.8 in 1965 to 142.4 in 1985, as farmers gave up the summer and winter cash crops and intercrops that had accounted for the high productivity per hectare-year during the late 1950s and early 1960s. The diversification index for individual farms declined from 3.0 in 1965 to 1.9 in 1982 as farm machinery use rose and farm labor dwindled. To meet market demand, the aggregate diversification index almost tripled between 1952 and 1985. According to census data, the full-time farm family decreased from 49.3 percent of total farm families in 1960 to 11.4 percent in 1985. Meanwhile, the remaining part-time farm families, with nonfarm jobs as their main employment, increased significantly.

Structure of Farm Family Income

Given that available farm family labor has been allocated to on-farm and off-farm activities, family income also derives from these two sources.

TABLE 6.6 The Multiple Cropping Index and Diversification Index in Taiwan

Year	*Cropping Index*	*Aggregate Diversification Index*	*Farm Diversification Index*
1946	117.2	NA	NA
1952	173.6	3.5	NA
1955	171.7	4.0	NA
1960	184.1	4.0	NA
1965	188.8	5.7	3.0
1970	183.0	6.7	3.1
1975	180.9	6.2	2.5
1980	154.4	8.4	2.1
1985	142.4	10.7	1.9

NA: Not available.

Source: Provincial Department of Agriculture and Forestry, *Agricultural Production Yearbook and Report on Farm Family Bookkeeping*, Taiwan, various years.

TABLE 6.7 Sources and Relative Farm Family Income[a]

Items	*1964*	*1970*	*1975*	*1980*	*1985*	*1987*
Structure (%)						
On-farm	64.7	48.7	46.3	29.6	36.7	38.1
Off-farm	35.3	51.3	53.7	70.4	63.3	61.9
Farm family income as % of nonfarm family income						
Per family	94.8	72.2	79.6	74.2	71.3	72.9
Per capita	69.8	60.2	76.1	66.4	69.0	72.6
Farm family income as % of laborer family income						
Per family	121.9	79.6	92.3	84.8	80.2	81.5
Per capita	88.1	70.0	81.5	79.9	83.5	87.3

[a]Laborers include production workers, and transport equipment operators and laborers.

Source: Taiwan Provincial Government, *Report on the Survey of Family Income and Expenditure*, 1987.

Off-farm income includes wages and salaries, sideline income, property income, and other miscellaneous income transfers. According to survey data, the annual increase in farm family income was attributable to rising off-farm income between 1964 and 1987. Real family income, on-farm income, and off-farm income increased annually by 4.2 percent, 1.8 percent, and 6.7 percent, respectively, during the same period. The growth rate and structure of farm family income from these two income sources define three stages between 1964 and 1987. Changes in the structure of farm family income paralleled those occurring in both economic development and agricultural policy strategies.

Stage I. Before 1970, earnings from farming were the major source of family income, accounting for between 52 percent (in 1969) and 65 percent (in 1964) of the total. As mentioned previously, agricultural production in the late 1960s fell into a price-cost squeeze dilemma due to unfavorable terms of trade. The widening gap of labor productivities between farm and nonfarm labor led to more extensive land use by an increasing number of farmers. On-farm income fell 5.3 percent annually from 1964 to 1970, while off-farm income increased by 5.7 percent per year. However, off-farm income rises could not offset on-farm income declines, so that real farm family income fell 0.7 percent annually. The gap between farm and nonfarm family incomes widened. The ratio of farm to nonfarm family incomes dropped from 94.8 percent in 1964 to 72.2 percent in 1970. Accordingly, the ratio of farm family income to worker family income decreased sharply from 121.9 percent to 79.6 percent, as shown in Table 6.7.

Stage II. Between 1970 and 1980, off-farm activities gradually came to play a more important role in increasing total family income because of the rapid growth of rural industrialization. The ratio of off-farm income to total family income increased from 51.3 percent in 1970 to 70.4 percent in 1980. On-farm income increased 2.1 percent per year, while off-farm increased 10.7 percent. In other words, off-farm income explained 89 percent of the annual 7.2 percent growth in total farm family income. However, the farm and nonfarm sectors were linked through markets for labor, capital, products, and current inputs. Therefore, the relative income ratio between farm and nonfarm families shifted erratically, depending upon both market conditions and government policies. In general, farm family income situations improved during this period.

Stage III. After 1981, the measures implemented in the 1970s produced gratifying increases in agricultural output, modernization and on-farm income. The proportion of on-farm income to total farm family income rose slightly from 36.7 percent in 1980 to 38.1 percent in 1987. Annual increases in on-farm income were much faster than those of off-farm income during this period. With declining competitiveness, Taiwan's agriculture has been hard hit by hefty agricultural imports. Shrinking farm product prices returned a flimsy profit margin, causing a decline in relative income for farm families in the early 1980s. By 1987, the ratio of farmer to nonfarmer income per capita was 72.6 percent.

The ratio of farm to nonfarm family income was substantially influenced by the business cycle in Taiwan. The larger the share of off-farm income in farm family income, the larger the income gap between farm and nonfarm families. This gap might imply that increased efficiency in farming is important to narrowing the income gap between the two sectors.

Farm Family Income Distribution

The distribution of farm family income is affected by patterns of on-farm and off-farm income. According to most literature, the high rate of economic growth and the spatially decentralized pattern of industrialization in Taiwan meant that off-farm sources of employment and income in rural areas contributed to the improvement of both income levels and income distribution among farm families. Therefore, to accelerate the process of rural development, stress should be placed on policies to support nonfarm activities, in accordance with the supply of infrastructure services in the rural areas (Ho, 1979; Chinn, 1979; Oshima, 1984). Both the inequality index of farm family income distribution (5th over the 1st income group) and the Gini coefficient suggest that income distribution in Taiwan improved with income growth from the 1960s to the mid-1970s, but became skewed with income growth since that date. In other words, the value of the Gini coefficient revealed a "U" shape in the process of economic development, as shown in Table 6.8, contrary to Kuznet's theorem.

In the earlier period of Taiwan's economic development, farm earnings were the major source of farm family income. The larger the farm holdings, the higher the farm family income. The distribution of farmland, therefore, directly affected the distribution of farm family income. The Gini coefficient of farmland distribution should be nearly equivalent to that of farm family income distribution. Given the size of farm holdings, the land/person ratio on small farms was less than on large farms. The small farms released unemployed manpower to nonfarm activities as opportunities arose. Therefore, the proportion of off-farm income was inversely related to farm size: the smaller the size of the farm, the higher the reliance on off-farm income. Thus, with off-farm employment opportunities, the distribution of income among different farm size groups become more even. The Gini coefficient fell from 0.31 in 1966 to 0.28 in 1975, and the 5th over the 1st income group inequality index fell from 5.16 to 4.12 in the same period. This indicates a considerable improvement in income distribution among farm families.

However, capital investment requirements for larger farms have increased sharply during the last decade. As farms were modernized, machinery and modern inputs increased. Labor demand fell on larger

TABLE 6.8 The Inequality of Farm Family Income

Year	*5th Income Group* / *1st Income Group*	*Gini Coefficient*
1966	5.16	0.31
1970	4.26	0.29
1975	4.12	0.28
1980	4.81	0.31
1985	4.52	0.33 (1982)
1987	5.01	NA

NA: Not available.
Source: Peng, T. K. (1987).

farms, whose excess workers found off-farm employment to increase their farm income. Off-farm income for larger farms has increased much faster than for smaller farms since the mid-1970s. Annual increases in off-farm income were 15.6 percent for farms less than 0.5 ha and 8.9 percent for farms greater than 2.0 ha from 1965 to 1968, but switched to 2.6 and 7.4 percent, respectively, from 1974 to 1982. The positive effect on the income distribution of increasing off-farm income for small farms was offset as the larger farms increased their own off-farm income through modernization. The Gini coefficient for farm family income increased from 0.28 in 1975 to 0.33 in 1982. The 5th over the 1st income group inequality index also increased from 4.12 in 1975 to 5.01 in 1987, indicating an unfavorable trend in income distribution among farm families.

In general, as the economy evolves toward high industrialization, the benefit of off-farm activities to farm family income may diminish, and farm family income distribution may become undesirable. Farm labor migration may become a more efficient and desirable strategy for small farms. According to the census data, the percentage of full-time to total farm households increased from 9.0 percent in 1980 to 11.4 percent in 1985. It appears that, as per hectare productivity growth slows with extensive small-scale mechanization, the growth of on-farm income tapers off. Thus, the extremely small part-time farm family may eventually migrate to nonagricultural employment.

Concluding Remarks

Under the family farm system in Taiwan, farm-level price signals allocated scarce resources among alternative uses, induced technological changes that substituted for relatively expensive resources, and dictated the level of on-farm income. Taiwan's agriculture, however, became highly dependent upon modern inputs. Factor prices were the primary determinant of on-farm incomes. Without exception, Taiwan's government, like every government in the world, directly and indirectly influenced farm prices and incomes in the pursuit of specific policy goals in the process of economic development. With two-thirds of farm family income consisting of off-farm income, the relative income problem for farm families remains.

To cope with further production cost increases and the widening income gap between farmers and nonfarmers, labor productivity in agriculture must be increased relative to that in nonfarm sectors. A policy of labor migration must be reinforced by publicly accepted programs developed to encourage part-time farmers to transfer their farmland and to assimilate farm people into the nonfarm population. For those who remain in farming, this can provide an opportunity to consolidate farm units of sufficient size to promote efficiency with modern machines and improved management. Also, such programs and policies could revise the regulations governing farmer's organizations, thus improving the bargaining power of farm groups and better representing farmers' political and economic interests. Further efforts should be made to increase the competitiveness of agricultural input markets, to change the product-mix, to promote farming efficiency, and to improve farm marketing channels. Moreover, throughout the transition period, a reasonable level of protection is both needed and justified, to facilitate the structural adjustment of Taiwan's agriculture.

References

Anderson, Kym. 1983. "Economic Growth, Comparative Advantage and Agricultural Trade of Pacific Rim Countries," *Review of Marketing and Agricultural Economics*, 51:231-248.

Anderson, Kym, and Y. Hayami. 1986. *The Political Economy of Agricultural Protection*, London: Allen & Unwin.

Binswanger, Hans. 1974. "A Cost Function Approach to the Measurement of Elasticities of Factor Demand and Elasticities of Substitution," *American Journal of Agricultural Economics*, 56:377-386.

Brown, L.R. 1977. "Population and Affluence: Growing Pressure on World Food Resources," in *Tradition and Dynamics in Small Farm Agriculture*, edited by R.D. Stevens, Ames: Iowa State University Press.

Chen, Y.E. and Y.T. Wang. 1980. *Secular Trends of Output, Inputs and Productivity*, The Institute of Economics, Academia Sinica.

Chen, Y.E. 1986. *An Economic Study on Agricultural Productivity in Taiwan*, Council of Agriculture (mimeograph).

Chinn, Dennis L. 1979. "Rural Poverty and the Structure of Farm Household Income in Developing Countries: Evidence from Taiwan," *EDDC*, 27:283-300.

Christensen, L.R. and W.H. Greene. 1976. "Economies of Scale in U.S. Electric Power Generation," *Journal of Political Economy*, 84:655-676.

Christensen, L.R., D.W. Jorgensen, and L.J. Lau. 1973. "Transcendental Logarithmic Production Frontiers," *Review of Economics and Statistics*, 65:28-45.

Council of Agriculture. 1986. *Long-Term Agricultural Development in the Republic of China (1986-2000)*, Taipei, Taiwan.

Ho, S.P.S. 1979. "Decentralized Industrialization and Rural Development: Evidence from Taiwan," *EDDC*, 28:77-96.

Johnson, D.G. 1973. *World Agriculture in Disarray*, Bongay: Richard Clay Ltd.

Lee T.H. 1961 "Taiwan's Farm Prices," *Journal of Bank of Taiwan*, 12:172-193. (in Chinese)

Meller, J.W. and B.F. Johnston. 1984. "The World Food Equation," *Journal of Economic Literature*, 22:531-574.

Ong, S.E. 1984. *The Development of Small Farms in Taiwan*, Taipei: Council of Agriculture.

Oshima, Harry T. 1984. *The Significance of Off-Farm Employment and Incomes Post-War East Asian Growth*, Asian Development Bank Economic Staff Paper, No. 21, Manila, Philippines.

Peng, T.K. and H.T. Chang. 1983. *Impact of Rapid Economic Structural Change on Agricultural Development in Countries with Small Farms*, The Case of Taiwan, ROC—The Unavoidable Productivity Lag, Taipei CAPD, June.

Peng, T.K. and M.Y. Chen. 1986. "A Study on Off-Farm Labor Supply and Labor Utilization of Farm Family in Taiwan," *Journal of Agricultural Economics*, 40:1-28. (in Chinese)

Peng, T.K. 1988. "The Change of Food Consumption Patterns-Implications to Rice Economy in Taiwan," *Journal of Agricultural Economics*, 43:69-96. (in Chinese).

Peng, T.K. and J.R. Lin. 1989. *An Economic Analysis of Input-Output Relationship for Taiwan's Agriculture*, Taichung: National Chung Hsing University, December. (in Chinese).

Timmer, C.P., W.P. Falcon, S.R. Pearson. 1983. *Food Policy Analysis*, Baltimore: The John Hopkin's University Press.

USDA. 1988. *Agriculture in the Uruguay Round*, ERS, December.

World Bank. 1986. *World Development Report*, Washington, D.C.: World Bank.

Yu, Y.H. 1973. "The Development of Diversified Farming in Taiwan," *Journal of Land Credit*, 10:1-14. (in Chinese).

Discussion

Guangsen Wang

It is well recognized that Taiwan has enjoyed remarkable industrial progress during the past few decades. Comparatively less has been known about the role that agriculture has played in this developmental process, but Tso-Kwei Peng's chapter has satisfactorily filled this gap. I will focus on reexamining some of the policy measures mentioned by Peng.

Agricultural price policy was accorded a strategic role in the process of Taiwan's economic development. Indeed, the government's switch from distortionary to supportive price policies traced the demarcation line between two phases of Taiwan's rural development. The use of price distortion as a measure to siphon agricultural surplus to nonagricultural sectors in the early stages of economic development is justified on the grounds that it is practically the only way to initiate industrialization for a late-developed economy. In Taiwan's case, I would add that population pressure and the unemployment problem in the early 1950s made the supply of cheap food grain even more critical to social security and political stability.

However, I would argue that a policy of price distortion is feasible only under two conditions. First, agricultural production must be promoted by other economic, financial, technological and institutional measures to offset the negative effects of price distortion. Secondly, there must prevail a considerable degree of self-sufficiency and isolation of the agricultural sector from the rest of the economy, making farmers less sensitive and accessible to the market. The second set of conditions may be found in most developing countries, where the economic behavior of the subsistence farmer is quite different from that of the agricultural entrepreneur in a mature commercial economy. The first set of conditions, though accomplished effectively by Taiwan in the early 1950s, were not always to be found in other developing regions. This difference explains why Taiwan stood out so spectacularly in its agricultural development in the 1950s and 1960s. Taiwan's experience, like Japan's, bears witness to the soundness of an agriculture-led growth strategy.

It is almost a matter of concensus among agricultural economists that a shift from price distortion to price support for agricultural products will take place sooner or later in the course of industrialization in a developing economy. This is what actually happened in Taiwan. The need for such a shift in price policy, however, was not simply or even primarily derived from a deterioration in the terms of trade for agricultural products. To the contrary, agricultural terms of trade had improved gradually in Taiwan before price support schemes were enacted in 1973 (Peng, Table 6.1). Nevertheless, with widening disparities in both rural-urban incomes and labor productivity between the agricultural and nonagricultural sectors, the unfavorable pricing of agricultural products became increasingly unendurable.

I do not mean to minimize the importance of price support systems, although some sources report that the average farm family benefited only moderately from price supports in Taiwan. Certainly, crop-specific price support can be highly effective in adjusting the comparative advantage of different crops or groups of crops; price stabilization schemes are generally beneficial to agricultural producers, in that they alleviate price hazards. But to rely upon price supports to protect the farm sector from rural-urban income disparities is not always desirable. Undue price support and government subsidy programs for the farm sector may help to ease the difficulties of farmers in the transition period, but the problems of intersectoral resource allocation remain unsolved. I share Theodore Schultz' perspective that, taking the long view, price supports may well be the worst of all policy options for reducing or eliminating this particular kind of income disparity.

I found especially interesting Peng's analysis of the contributions of increased inputs and technical change to the growth rate of agricultural output (Peng, Table 6.4). However, I would like to summarize the findings from this exercise somewhat differently:

1. For the three decades from 1952 to 1981, agricultural production in Taiwan was enhanced mainly by the increase in conventional inputs, which accounted for nearly 60 percent of the annual growth rate of total output.
2. The rates of increase for the different categories of inputs are inversely associated with changes in their relative prices; thus, the input structure of agricultural production varies in the different stages of development.
3. In the later stage of development, labor input actually decreased at an accelerating rate, leaving capital inputs solely responsive to the increasing pressure of agricultural growth.
4. Both the growing share of input increases in the growth rate of agricultural output and the overwhelming role played by current capital input in the later stage suggest that the potential of traditional technology had been exhausted. Actually, the growth rate of agricultural output had been reduced to the margin of stagnation (barely 1.1 percent) by 1977-81. To revitalize agricultural growth, there was an obvious need to seek breakthroughs in biological, land-saving, and scale-neutral technologies geared to increasing land productivity, and to introduce mechanical technology geared to saving labor.

Incidentally, I would like to point out that the decreasing trend of farmland use (in terms of crop area) is a sign of the extensive use of land only with respect to labor intensity. The multiple cropping index could well be taken as an appropriate indicator of land-use intensity so long as labor remained the dominate factor in agricultural production. But in present-day Taiwan, field crops constitute only a minor part of agricultural output, while labor has been increasingly substituted by capital inputs. Agriculture has thus become capital-intensive rather than labor-intensive. In this context, the multiple cropping index has lost much of its original meaning as a measure of land use intensity. (The same can be said of many terms in the glossaries of our agricultural economics textbooks.)

Rural development in Taiwan is characterized by a large-scale outflow of farm labor and widespread part-time farming. Outflow of farm family members to seek jobs in the city is a time-honored practice in Taiwan, but the pace of outflow accelerated rapidly in the 1960s. Consequently, the proportion of the farm population to the total population declined from 40 percent in 1970 to 20 percent in 1987. Although the number of farm *households* showed little variation, the number of *persons per family* decreased from 7.2 to 4.5. Hence, average farm size remained at about 1 ha, but cultivated land per farm laborer increased considerably.

Among Taiwan's 771,906 farm families in 1985, 683,905 were part-time farm households. In other words, nearly nine-tenths of the farm households in Taiwan in 1985 derived their family income partly or mainly from off-farm income sources. I would like to stress the importance of differentiating these part-time farm households into two categories: those earning their family income (1) chiefly from farming, and (2) chiefly from other jobs. Resource allocation inclinations differ markedly between these two categories of farm households. Generally, part-time farm households in category 1 are ready to direct their resources toward upgrading land productivity. Part-time farm households in category 2 are more reluctant to invest in farmland improvement, because they take farming only as a sideline business. It is noteworthy that these sideline farm households in Taiwan increased persistently from 153,442 in 1960 to 530,712 in 1985. That is to say, 68.8 percent of all farm households in Taiwan in 1985 were farm households that took farming as their sideline business.

In light of the major influence of off-farm income on the income distribution among farm families, I do not know how relevant the Taiwan experience of a "U" shaped movement of the value of the Gini coefficient is to other developing regions. If Peng's notion that "the proportion of off-farm income was inversely related to the farm size, the smaller the size of farm, the higher the reliance on off-farm income" is correct (I tend to think so), then the statement that "as the economy evolves toward a high industrialization economy, the benefit of off-farm activities to farm family income may diminish, and farm family income distribution may become undesirable" should be scrutinized more carefully and fuller explanation should be made. Elsewhere, regarding the performance of farm family income against nonfarm family income between 1964 and 1987, Peng states that "the larger the share of off-farm income in farm family income, the larger the income gap between farm and nonfarm families." I am not clear whether this statement means that smaller farms with higher reliance on off-farm income were worse off than larger farms as compared with nonfarm families, or that farm families as a whole lagged further behind nonfarm families in their income levels as the share of off-farm income became larger and larger. However, basing my observation upon the empirical evidence of Taiwan and other developing regions with a high man-land ratio, I would speculate that:

1. Off-farm income would contribute substantially to the enhancement of farm family income, and would tend to make the income distribution among farm families more even and the gap between farm families and nonfarm families smaller.
2. Many other factors contribute to income variations; hence, off-farm income should not be taken as the only or dominant factor in explaining these income variations.
3. Off-farm income could not reverse the declining trend of the farm sector in a rapidly growing economy. However, part-time farming, if wisely guided, might serve as a stepping stone in the structural transformation of the agricultural sector.

I wholly agree with the author in that the basic structural problem in Taiwan's agriculture is the extremely small size of the farm operation. As the above discussion reveals, although price supports and off-farm income could be beneficial in lessening the plight of farm families in times of growing rural-urban discrepancy, these measures are not tailored to raising agricultural productivity. Such other measures as land consolidation, joint operations, entrusted farming, and co-operative farming were designed to enable small farms to perform with large-farm efficiency, but the income level of the individual farm family still lags far behind that of the nonfarm family under such arrangements. In any case, average farm size must be enlarged if agricultural productivity is to be greatly enhanced. This was exactly what the second stage of the land reform program attempted in Taiwan.

Peng advocates in his concluding remarks that a policy of labor migration must be publicly accepted; programs must be developed both to encourage part-time farmers to transfer their farmland and to assimilate farm people into the nonfarm population. Clearly, very serious institutional, social, political, economic, and ideological difficulties arising from the private ownership of land must be confronted if such a policy suggestion is to be implemented. I would not impose on anyone the advantages of the collective ownership of land; indeed, many theoretical as well as practical problems need to be solved for a system of collective land ownership to function well. But I believe that an objective, sober-minded, and free discussion would be beneficial and fruitful.

7

Prices, Costs, and Farm Income in Mainland China: Rules of Farmer Behavior in a Uniquely Mixed Economy

An Xi-Ji

The interrelations among agricultural prices, production costs, and farm income constitute one of the most urgent economic issues facing agricultural economists today, whether in China or abroad. Applied research on the subject in recent years has yielded divergent findings and points of view. Rather than adding to this already-rich empirical experience, the present chapter attempts to focus on how the economic mechanism underlying price-, cost- and income-movements operates to guide farmer behavior in the uniquely mixed economy of Mainland China. By tracing how the farm sector has navigated its way through the economic reforms of the 1980s, we shall endeavor to shed light on agricultural practices in China.

Instead of constructing an overall model, this chapter will address economic processes one by one. Past work on this subject has, in general, systematically stressed neither the workings of the economic mechanism nor the rules of producer behavior. Exceptions include Tian, whose work covers a scope similar to ours; and Sicular, who analyzes the interaction between plan and market in China's agricultural trade from a theoretical standpoint.

This chapter will be divided into four parts. The first will outline current agricultural practices in China, as well as the theoretical framework we shall use to analyze them. The second will describe the farmers' decision-making process. The third section will address resource-use efficiency. Finally, we shall offer some perspectives on the sustainability of the mixed economy.

Practices and Theoretical Framework

The economic system of rural Mainland China in the 1980's is a unique mixture of four distinct elements linked by mutual interactions. In addition to the market- and planned economies, there exist the *control economy* and the *self-sufficiency economy*. This unique mixture has resulted from economic reforms in both land-use and commercial systems during the 1980's.

The household responsibility system (HRS) emerged in 1979 and the early 1980s. In essence, the system resulted in a land reform. There are two main features of the new system. The first is the transfer of land management rights from collectives to individual households, even though the rights of transfer among households were restricted and land ownerships were kept the same. The central government announced in 1983 that households would have the right to farm the land distributed to them under the HRS for 15 years. What is not known to farmers is when and how the collectives or the local government will make adjustments in this land distribution.

The second main feature of the new land system is that the collectives retain land *ownership*. Thus, the farmer's status has been changed from commune member to independent co-proprietor, but he is still linked by contracts to the collective. In economic terms, this means that while the households manage their farms as income-maximizers, they must both consider the supply of land as fixed and deliver procurement quotas to the state at low prices. Collective land ownership has come to be considered a political and ideological token of socialism in China; but at the same time it has been used as a basic instrument for economic planning.

In the sphere of commercial reform, three measures have been implemented in the 1980s:

1. *Gradual loosening of the state's trade monopoly*. This has occured both in scope and in strength. For example, by 1984, the number of agricultural commodities under state procurement had been slashed from 180 to just 21. In addition, the size of compulsory quota deliveries at low price was reduced between 1979 and 1984, but raised thereafter for grain and cotton.
2. *Creation and development of free markets*. This has brought into play partial market mechanisms in both the countryside and the cities. But free markets have remained to varying degrees under the control of state commerce, and thus subject to frequent administrative interference in such forms as boundary blocking.
3. *Restructuring of state commercial agencies*. In the original planned economy, state commerce monopolized the market in accordance with a unified overall national plan; state commercial agents operated on the

order of administrative officials without assuming responsibility for profit and loss. All expenditures and revenues of these agents had to be submitted to the national treasury. The avowed aim of the commercial reform was to establish a new type of state commercial agents who, like independent businesses, would be responsible for profits and losses. These agents should compete with, rather than control, private business in the free market. At the same time, they should fulfill national plan targets set from above.

This administrative restructuring has occured mainly since 1985, and includes such measures as decentralized decision-making; new bonus systems; and various forms of responsibility system, such as contractual arrangements, that make the manager responsible for a certain amount of profit delivery. Two major results have flowed from this restructuring. First, a multichannel market has emerged, with state commerce enjoying a semi-monopoly. Second, state commercial agents have adopted new positions of being independent business organizations responsible for profit and loss.

The structure of the multichannel market is subject to change every year, but may be roughly described under four headings:

1. Total state monopoly of cotton markets.
2. Compulsory contract quotas in two-thirds of grain trade, with the remainder sold on the free market.
3. State commercial control of 70-80 percent of meat, poultry, eggs and milk markets in big cities, and 50 percent in small and medium cities, with rural areas left to the free market.
4. Monopoly by one specific agent of agricultural input marketing (e.g., fertilizers, insecticides, diesel oil, plastic film). In fact, the Supply and Marketing Cooperatives are the second most important national commercial ministry active in the countryside.

The new state commercial agents have essentially taken on the characteristics of semi-independent businessmen and semi-administrative planners. They merge plan and market into one, simultaneously carrying out economic plans while pursuing their own benefit. These agents, most of whom have emerged since 1985, constitute the main element of the control economy. Their marketing activities seriously affect agricultural prices, production costs, and farm incomes.

It goes without saying that self-sufficiency directly influences trade structure and crop output composition, and hence the level of resource-use efficiency. But self-sufficiency is not directly involved in the movements of prices, costs, and cash farm income.

Farmer Decision Making

Is agricultural procurement a tax or a business between the government and farmers? Whatever the answer, such procurement remains a vague notion in economic circles in China. I will clarify the concept here, before going into farmer decision-making. My personal opinion is that agricultural procurement is a tax. More specifically, it is a product tax in kind, because both the amount and the "price" of procurements are unilaterally decided by the government. For farmers, procurements are compulsory, without any room for bargaining. Stalin, in the late 1920, was the first person in socialist countries to use the term "price scissors," in the sense of the difference between the prices of agricultural products in the free market and by procurement quota. He called procurements "an extra tax on the farmers." The experience of the last 40 years in China shows that Stalin was right theoretically, even though his way of expressing himself was unclear. The extra tax is based upon the total output by each household of a specific product, such as grain, cotton, and hogs. These must be delivered to the state in kind, usually at a much lower price than in open (or black) markets. The *amount* of the tax in value equals

$$A^* (P - P')$$

and the tax rate is A/T in physical terms, or $A^*(P-P')/PT$ in value terms, where

A = the amount procured
P = the market price
P' = the procurement price
T = the total output of a given commodity.

It is obvious that the rate of the extra tax is subject to fluctuations in A, P', and T, given the level of market price. A and P' are both determined by the government, as mentioned above, while T is decided by farmers, at least to some extent.

Let us turn now to the process of farmer decision-making in the uniquely mixed economy just described. Our approach is based on five assumptions:

1. Farmers manage their farms with a view to profit maximization, and respond appropriately to market signals.
2. Both self-sufficiency and procurement quotas are strictly rigid.
3. Market price is not affected by changes in the product tax. In a market economy, an additional product tax will shift the supply curve backward, subsequently raising the price level. In the uniquely mixed

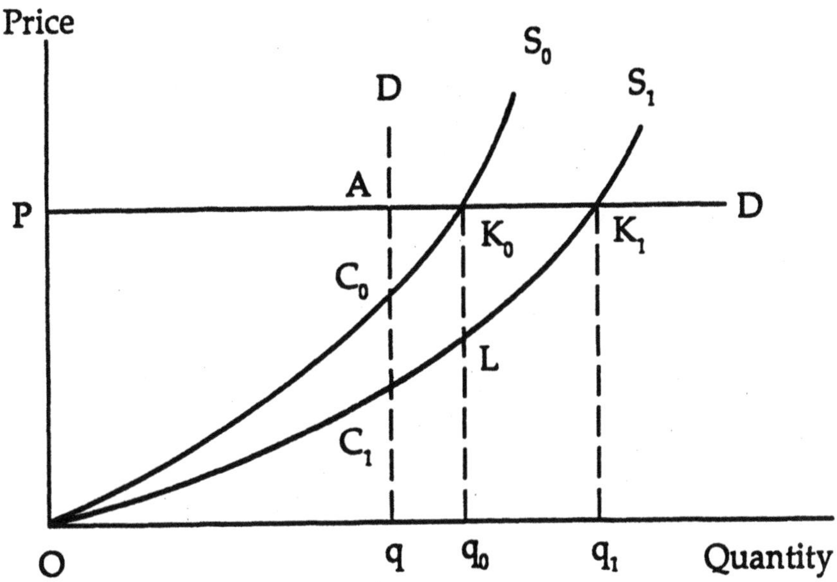

FIGURE 7.1 Price, Cost, and Farm Income

economy, the product tax would shift the supply curve backward, just as in a market economy (i.e., from S_1 to S_0 in Figure 7.1). However, three factors make it unclear how a product tax interacts with market price. First, the tax is paid in kind. Second, the government will subsequently allocate the procured commodities to the inhabitants of cities and towns at even lower subsidized prices than the procurement price P′. And third, these urban-dwellers constitute the main part of domestic demand for agricutural commodities. I assume here that the burden of the product tax falls totally on the farmer, excluding the possibility of transferring all or part of the tax to the consumer.

4. International trade has no effect on price.
5. Production volume and composition are given.

The optimal (or nearly optimal) total output in Figure 7.1 is Oq_0, corresponding to the intersection of P and S_0, in contrast to the output Oq_1 without a product tax.

It is obvious that both total output and net cash income are sensitive to changes in the product tax. If the government reduced the rate of product tax by decreasing the amount of quotas and/or by raising the quota "price" (P′) paid to farmers, the supply curve S_0 would shift outward, thereby enlarging both output and the farmers' net cash income.

This is what happened in the period 1979-1984. For example, planned quota deliveries of grain decreased 12 percent from 1978 to 1981, and the grain quota actually delivered by farmers declined 36 percent in the same period. At the same time, the price paid to farmers in 1979 for grain quota delivered increased 20 percent, for cotton 15 percent, for hogs 26 percent and for other agricultural quotas by from 20 percent to 50 percent. Between 1979 and 1984, additional increases in quota prices paid to farmers were announced every year, with specific amounts of change varying from year to year. In 1983 and 1984, the prices paid to farmers for quota deliveries of grain were almost the same as those in the free market. The supply curve with product tax S_0 almost coincides with the supply curve without product tax S_1.

On the other hand, if the government raises the product tax by increasing the amount of quota procurements and/or by dropping the procurement "price" (P′), the supply curve S_0 shifts backward. The result is decreased output and an erosion of the farmers' cash net income. This is precisely what has happened for grain and cotton since 1985, when grain quota procurements began to rise swiftly. The average annual grain quota procurement in the period 1985-1987 was double that for the period 1980-1984. While market price levels for grain crops rose 41 percent in 1984-1987, the procurement price paid to farmers rose by only 7 percent. To cope with this heavy product tax, farmers transferred their productive resources from grain and cotton to nonagricultural trade and/or the production of agricultural commodities with lower or zero product taxes.

Agricultural input prices, decided by semi-independent state monopoly agents, affect the levels of both output and cash net farm income through their influence on production costs.

Under a given product tax and the same assumptions as previously, production costs, output level, and cash net farm income are all subject to changes in input price levels. If the monopoly agents implement a policy of reducing input prices, the supply curve will shift outward from S to S′ (Figure 7.2). This implies that marginal cost declines, output quantity rises by qq′, and cash net farm income grows by CKK′C′. This is what happened in the first half of the 1980s. If, instead, monopoly agents had raised input prices, the supply curve would have shifted backward from S to S″, thus raising marginal costs, reducing output by q″q, and cutting cash net farm income by K″KCC″. This latter scenario has occurred since 1985. Numerically, the average mixed price of fertilizer rose 33 percent from 1984 to 1987, while grain output price increased by only 7 percent.

Many Chinese economists speak frequently of the "twists" in agricultural prices, production costs, and farm income in China, in the sense of deviations from the rules of a market economy. It is obvious from the above paragraphs that all such twists arose from the workings of a mixed economy and the "normal" reactions of farmers thereto.

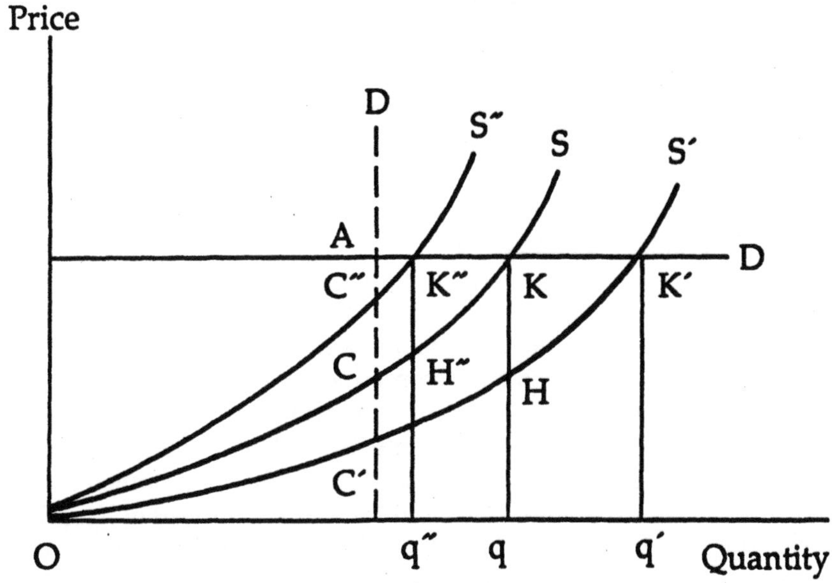

FIGURE 7.2 Changes in Input Prices

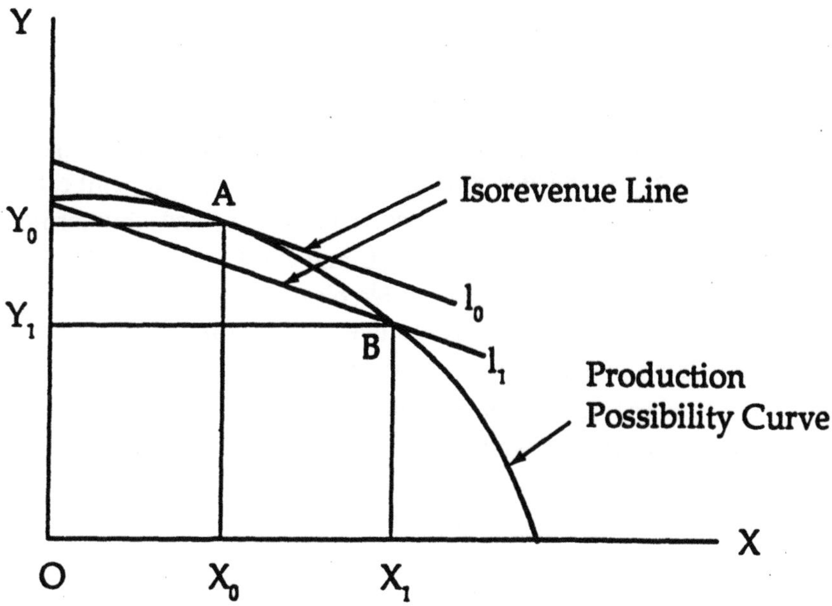

FIGURE 7.3 Combination of Two Products X and Y

Resource Use Efficiency

In the mixed economy observed in China, resource-use is inherently inefficient. This inefficiency stems primarily from the misallocation of variable inputs, the nontransferability of land-use rights among households, and government subsidies to agriculture in the form of parity prices on variable inputs. The following example may explain why resources are misallocated under a quota system.

Suppose a farmer selects an output combination of two products, X and Y, but has limited access to variable inputs (Figure 7.3, Tian). In a market economy, the optimal output combination would be at point A, where the isorevenue line (I_0) is tangent to the production possibility curve. The farmer would produce X_0 of X and Y_0 of Y, thus earning the greatest revenue possible (I_0) for his limited variable inputs. By contrast, if the quota system fixes his compulsory deliveries of X at $X_1 > X_0$, then the farmer would rationally produce at point B. He would produce X_1 instead of X_0 and Y_1 instead of Y_0, making total income of $I_1 < I_0$. The farmer suffers a loss, as does the agricultural sector in general.

Under the HRS, the average household has about 0.6 ha of arable land, scattered among 5 to 9 fragments around the village. Farm size is "super-small." As indicated, land use rights are not transferable among households. Such a land system inevitably leads to inefficient farming.

Imagine that two farmers, A and B, possess differing amounts of variable inputs and technology. AA′ in Figure 7.4 (Tian) represents the value marginal product curve of A's land ($VMPL_A$), while BB′ depicts the value marginal product curve of B's land ($VMPL_B$).

In a market economy, the distribution of land between farmers A and B should theoretically stabilize at $VMPL_A = VMPLA_B$. This is at point c, the intersection of AA′ and BB′, where farmer A controls L_bL_1 land while farmer B controls L_aL_1, earning revenues of cAL_bL_1 and cBL_aL_1, respectively. But under the HRS— with land non-transferable—land would be equally distributed between A and B, irrespective of the farmers' capacity to produce marginal value. Farmer A is allocated land L_bL_0 and B receives L_aL_0 with revenues of aAL_bL_0 and bBL_aL_0, respectively. The loss of net revenue is given by the triangle abc, where

$$abc = (cAL_bL_1 + cBL_aL_1) - (aAL_bL_0 + bBL_aL_0)$$

Furthermore, because land rights are not legally guaranteed for a sufficiently long period, farmers hesitate to invest in land improvement projects. The perfectly natural result is a continuous reduction in soil fertility. (This problem is beyond the scope of the present paper.)

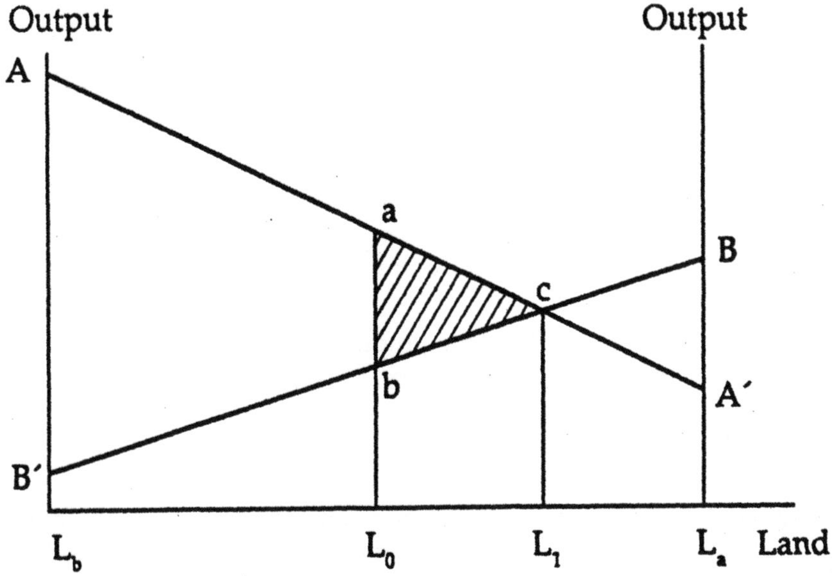

FIGURE 7.4 Land Use Under Household Responsibility System

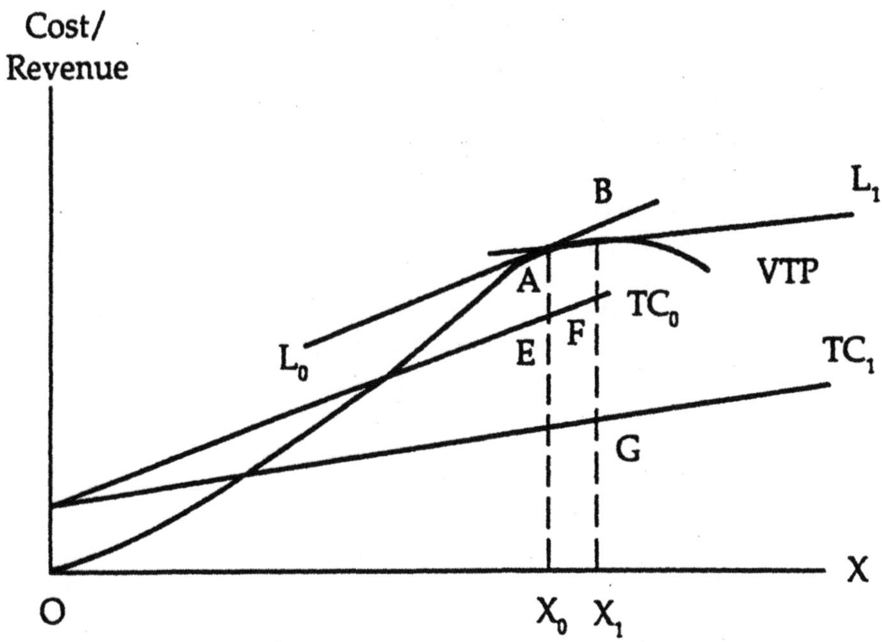

FIGURE 7.5 Optimum Amount of Input With and Without Subsidized Input

The optimal amount of input, using the value of the total product curve VTP and the total cost (TC = TFC + P*X), is determined by the following equation in a market economy:

$$P_Y * MPP = P_X \, , \text{ or } VMP = P_X$$

where P_Y and P_X are the prices of the product Y and the input X, respectively.

At the point A (Figure 7.5), VTP is tangent to L_0(//TC_0), and input X amounts to X_0. Profit is AE, the maximum possible given the production function, P_Y and P_X.

If the government subsidizes the production of a specific product by supplying variable inputs at below-market prices, TC would shift downard from TC_0 to TC_1. The new optimal amount of input X would be X_1, corresponding to point B, where VTP is tangent to L_1(//TC_1). Profit is BG = BF + FG. FG represents transfer income to farmers from nonagricultural sectors. But AE > BF, so in terms of production efficiency, the non-subsidized case is preferable to the subsidy.

Perspectives —The Sustainability of the Mixed Economy

Notwithstanding the presence of twisted prices, costs, and farm incomes, and the inefficiency of resource use, the mixed economy observed in China may be sustainable with slow growth of the national economy if plan evasion can be contained within a limited scope.

Plan evasion, which seems unavoidable given China's experiences over the last 30 years, can take at least three forms. First, if procurement prices of different products are out of balance in terms of the farmers' benefit, farmers are induced to transfer their resources from the "inferior" product to the "superior" product, even under the unified planning economy. Such a shift occured in the 1960s and the 1970s between cotton (inferior product) and grain (superior product). For almost two decades (1960-1978), procurement quotas for cotton went under-fulfilled in North China, one of the two main cotton areas. By 1978, the average yield of cotton had plunged to less than 300 kg/ha in Shandong and Hebei provinces. In the first half of the 1980s, the wind shifted in favor of cotton and other economic crops against grain; farmers responded by converting land from grain to economic crops.

A second form of plan evasion occurs in the presence of a free market. Lured by higher prices, farmers transfer part of their procurement quota to the market.

After the emergence of semi-independent state commercial agents in 1985, still a third form of plan evasion has occurred in China. Although formally charged with the mere implementation of the national plan, these agents also pursue their own stakes in the course of exchanging agricultural products (procurements) and agriculutral inputs (subsidy) with farmers. Attracted by personal benefit, agents evade plans by dropping procurement prices (including paying nothing for grain quota to farmers,) and raising input prices by transferring subsidized inputs to the free market.

From the farmers' side, plan evasion usually results in a deviation from the income distribution intended by the plan. From the commercial agents' side, it often results in outright inequity, and is much more difficult to deal with.

Another inherent characteristic of the mixed economy is the instability of widely-fluctuating agricultural prices. Farmers cannot, in general, get appropriate price information in time; the product tax rate is irregularly decided by officials to meet the specific goals of the moment. Other things equal, the greater the quota deliveries demanded and the more rigid the delivery system, the less the market supply and the higher the price. Price level fluctuates with changes in both government policies and the farmers' supply response.

Conclusion

In conclusion, whether China's uniquely mixed economy is economically and socially sustainable may depend mainly upon whether semi-independent state commerce is a transitional or a durable phenomenon. As the central authorities have once again conceded, the only solution to China's current marketing problems is further reform. The most pressing need is to find effective and enforceable devices to restrain the rather popular practice of plan evasion, especially on the part of semi-independent state commercial agents.

References

Sicular, Terry. 1988. "Plan and Markets in China's Agricultural Commerce," *Journal of Political Economy*, Vol. 96, No. 21.

Stalin, J. 1929. *Works of J. Stalin*, Vol. 12, p. 45 (in Chinese).

Tian, Wei-Min. 1980. "The Surplus Product Market Economy," unpublished paper in Chinese, February.

Discussion

Carson Kung-Hsien Wu

It is both pleasurable and fruitful to read Professors Peng's and An's papers (Chapters 6 and 7) on the issues of agricultural prices, costs, and income on each side of the Taiwan Strait. I will first summarize their findings; then discuss the presence of a mixed economy on both sides of the Strait, each with different constraints to further development; and finally provide some suggestions for both Taiwan and Mainland China.

Professor An's paper covers three points:

1. *Emergence of a uniquely Chinese mixed economy.* This mixed economy has been formulated by introducing two reforms since 1979, especially after 1985. It consists of four distinct elements: the planned economy, the self-sufficiency economy, the market economy, and the control economy, of which the last two resulted from two reforms. These were (1) the adoption of the household production responsibility system, which is itself derived from the land reform; and (2) the establishment of "new" forms of state agents in the commerce (or circulation) system. The new forms here are (1) multichannel marketing with semi-monopoly of state commerce, and (2) state commercial agents with semi-independent business and semi-administrative planning characteristics.
2. *Impact of these two reforms on both farmer decision-making behavior and efficiency resource use.* Professor An uses the market mechanism to explain farmer behavior in consideration of procurement prices and amounts, production costs, and farm income. He concludes that both the "abnormal" behavior of the farmers and the mechanism of the mixed economy twisted prices, production costs, and farm income away from the rules of a market economy. He also concludes that resource use is inherently inefficient, due to the misallocation of resources and government interference in the form of quota and land tenure systems of this Chinese mixed economy.
3. *Challenges in the future.* Professor An concludes that the planned economy segment will unavoidably grow as unstable agricultural prices become one of the inherent characteristics of this Chinese mixed economy. Furthermore, he stresses that "the most pressing need is to find effective and enforceable devices to restrain the rather popular practice of plan evasion, especially on the part of semi-independent state commercial agents."

Professor Peng's paper also details three items:

1. *Price situation in Taiwan.* This section covers various price policies for an evolving agricultural context. Government programs include price supports, price stabilization, direct payments, and input subsidies to maintain agricultural prices at reasonable levels and to reduce both production costs and the financial burden of farmers.
2. *Changing agricultural structure.* Professor Peng discusses (1) input combinations, (2) output composition, (3) productivity, and (4) agricultural trade. Taiwan's agriculture has become more market-oriented, commercialized, industrialized, and internationalized. Technologies have played a major role in the development process.
3. *Dynamics of farm family income.* On-farm and off-farm income sources have exchanged their relative positions due to the development of rural industrialization and the migration of agricultural labor. The ratio of income from on-farm and off-farm sources was 65:35 in 1964, but dropped to 38:62 in 1987.

Comments and Supplements

A Mixed Economy Prevails on Both Sides of the Taiwan Strait. Fundamentally, the government on Mainland China has generally disregarded the factors of production costs, price incentives, and management techniques in agricultural production, especially that of food grains. The operation of the agricultural sector has been mainly controlled by the central government with a system of food and income rationing according to the principles of need, labor (or work), and egalitarianism of the reward system. Unfortunately, this egalitarianism often degenerated into the under-motivation of equalitarianism. Only for the past 8 to 10 years, in particular since 1985, have the procurement prices of rationed goods been raised, the rationed amounts relaxed, the spread between basic and incentive wage levels enlarged, and cropping systems adjusted to local conditions. The agricultural economy has been shifted from orthodox socialism toward greater liberalism.

Farm income increased from 40.5 *yuan* in 1957 to 52.3 *yuan* in 1965, and from 62.8 *yuan* in 1976 to 85.9 *yuan* in 1980 (Kao, p. 25). In addition, Johnson (1988) has estimated that the real income of farm families approximately doubled between 1978 and 1986. Everyone technically had a job, although disguised unemployment (surplus labor) was estimated at 127.9 million people in 1981 (Kuo, pp. 35 and 37).

Throughout Chinese history, a given regime has always run the risk of being overthrown if things came to be intolerable even to the docile farmers of rural areas. At present, 80 percent of the population in Mainland China remains rural, relying mostly on agricultural activities for their income and employment. Perforce, agriculture has been treated by the regime as critically important to maintaining its internal stability and international position. Income increases, though minute, did help stabilize the regime on Mainland China. In spite of remarkable progress by statistical measures, *de facto* income and consumption levels have remained low by world standards.

On the other hand, Taiwan adopted price distortion and taxation approaches to siphon agricultural surpluses to nonfarm sectors in the early stages of economic development. The production and marketing of rice, the principal food grain for not only domestic consumption but also foreign exchange in 1950-1964, has been tightly controlled. Yet the structure of agriculture has undergone drastic changes, and farm family income has derived from various sources. Agriculture has developed to fully use the available resources.

Differing Constraints to Further Development. The leaders on Mainland China have admitted that their policy mistakes discouraged the positive attitude of Chinese farmers in agricultural production and lowered productivity due to lack of willingness to work (Kao, pp. 13-14). Government policies had misled people from efficient and effective allocation of resources (Chang, p. 69). These measures resulted in both stagnant or declining productivity and a deteriorating natural environment (Tien). Moreover, the cultural characteristics of the Chinese people have influenced development. Traditionally, peace-loving and diligent Chinese farmers living in harmony with nature showed little concern for the ruling class. Whenever the authorities became authoritarian, totalitarian, or irrational, farmers took the attitude of passive resistance; yet they have responded actively and positively to reasonable policies, just as all peasants around the world (Thiesenhusen). This illustrates why the Mainland government has failed to reap a genuine harvest of agricultural development, even with successful control over the people.

Economic problems can only be solved by economic methods. The better production performance on Mainland China during the periods 1949-57 averaging 14.1 percent annual growth rate and 1980-88 (at 10 percent)[1] (Yeh, pp. 38-39) can be attributed to economies of scale and economic incentives. Since 1979, the adoption of the household production responsibility system has provided the incentives for farmers to efficiently use the resources at their disposal and to share the results of their efforts within a reasonable distribution system. The production growth rate once again climbed to two digits (Tien, p. 46). Recently, agricultural growth has slowed; nevertheless, the diversification of products has helped to rebalance the ecosystem. This diversification has resulted in a more reasonable product balance over the past 10 years (1979-88). While forestry, livestock, and aquatic products remained at 3 percent, 13 percent, and 1.3 percent of output value, cultivated crops decreased to under 60 percent within crop production. The share of food grains declined within crop production, while cash crops (e.g., sugar) and feed grains increased. Meanwhile, sideline activities rose to 30 percent, of which rural industries occupied an increasingly overwhelming proportion (60 percent in 1971, and 90 percent in 1984) (Lay, p. 11).

In Taiwan, the agriculture sector has made its contribution to economic development and adjusted itself to changing patterns of domestic consumption and international agricultural trade. Nevertheless, small farm size and an abundant labor force in agriculture remain major constraints to future development. Furthermore, environmental protection, in terms of both the operation of and the impact on agriculture, constitutes the critical social issue for future development.

Taiwan's Experiences Can Be Applied for Further Improvement of Agricultural Development on Mainland China. This experience may be applied in the following aspects:

1. *Full utilization of labor and improvement of labor quality through a well-established education system.* In the early stages of agricultural development, farm laborers in Taiwan were fully employed throughout the year by working not only on their own farms through multiple cropping systems, but also cooperatively on such public works as irrigation, transportation, and community development projects. These works were managed by a *counterpart system,* in which the government bore an equal share of the expense with the farmers. The government provided material and management skills, while farmers contributed labor. Both vocational schools of agriculture and their graduates working in farmers' associations contributed significantly to agricultural development in Taiwan. The creativity of farmers has been the major factor in the dynamic adjustment of Taiwan's agricultural development.
2. *Dynamic forces in rural Taiwan.* Three forces have facilitated agricultural development in Taiwan. These are:
   a. the substitution of motorcycles for bicycles to mobilize the labor force in rural and urban areas;
   b. the development of rural industrialization to make full use of the abundant labor force, to distribute agricultural production, and to promote cooperation between agriculture and industry; and
   c. the urbanization of rural areas in the form of rural electrification,

community development, and a well-paved transportation network with the implementation of land consolidation. In contrast, for example, motorcycles, the only efficient transportation in mountainous areas and the narrow alleys of urban centers, are commonly lacking in Mainland China.

3. *Expert assistance.* Agricultural development in Taiwan has been virtually guided by experts. As Professor Mao has demonstrated in Chapter 5, the personnel of the Sino-American Joint Commission on Rural Reconstruction (JCRR) and the graduates of vocational schools have been in active service in rural areas.

   On the other hand, on Mainland China, communes have been ruled by government officials, veterans, or cadres who joined the Long March of the 1930s. As one Chinese saying goes, "Those who retire from the army may well return to the farms." This may have been true for traditional society and production techniques. Unfortunately, modern agriculture can no longer be guided by persons with little perception of advanced technology and theories.

Conclusions

In sum, there is room for improvement in both Mainland China and Taiwan. The potential for increased production throughout China is there, but how can it be realized? Existing production and distribution systems need further modification. On the one hand, economics, especially agricultural economics, can offer much in accelerating the agricultural development on Mainland China. On the other hand, the functions of cooperatives in Taiwan can be strengthened to reap economies of size in both farm operation and agricultural marketing. Chinese farmers in both the Mainland provinces and Taiwan are hard-working and obedient. They have in the past, and will continue in the future, to make the greatest possible contribution to their country's development that is consistent with wise policies and reasonable motivations. The challenge is how best to channel their intrinsic diligence!

Notes

1. The recent revised rate was 6 percent, of which 7.7 percent for the subperiod 1979-1984 and 4.0 percent for the subperiod 1985-1988 (Johnson, 1989).

References

Chang, Pei-Chen. 1982. *A Study of Regional Factor Productivities of Agriculture in Mainland China* (in Chinese), Economic Papers No. 9, Chung-Hua Institution for Economic Research, Taipei City, ROC, August.

Johnson, Glenn L. 1988. "Institutional Reform: Accomplishments and Unmet Needs in China, Newly Industrialized Countries of Asia, the Soviet Union and Eastern Europe and the Developed Countries," *Conference on Directions and Strategies of Agricultural Development in the Asia-Pacific Region*, pp. 9-43, January 5-7, The Institute of Economics, Academia Sinica, Taipei City, ROC.

Johnson, D. Gale. 1989. "Economic vs. Noneconomic Factors in Chinese Rural Development," Paper No. 89:13, June 19, Office of Agricultural Economic Research, University of Chicago, 23 pp.

Kao, Chang. 1983. *An Estimation of Labor Surplus in Rural Mainland China* (in Chinese), Economic Paper No. 38, Chung-Hua Institution for Economic Research, Taipei City, ROC, September.

Lay, Tzyy-Jane. 1988. *The Economical Analysis of Agricultural Structure Change in Mainland China* (in Chinese), Economic Paper No. 115, Chung-Hua Institution for Economic Research, Taipei City, ROC, March, 90 pp.

Thiesenhusen, William C. 1989. "The Economic Contributions of Peasants are Underestimated," *Choices* 4 (2) : 30-31, Second Quarter, American Agricultural Economics Association, USA.

Tien, Jiun-Mei. 1984. *A Study of Mainland China's Cropping System* (in Chinese), Economic Paper No. 53, Chung-Hua Institution for Economic Research, Taipei City, ROC, August.

Yeh, Shin-Hsing. 1985. *An Anatomy of the Production Responsibility System in Agriculture in Mainland China* (in Chinese), Economic Paper No. 57, Chung-Hua Institution for Economic Research, Taipei City, ROC, March.

PART FOUR

Trade and Competiton

Cut flowers are a labor-intensive, high-valued crop that is booming in Taiwan. *(Courtesy Council of Agriculture)*

Introduction to Part Four

Wen S. Chern

Chapters 8 and 9 deal with trade and competition. Mainland China and Taiwan represent interesting contrasts in their agricultural trade. Mainland China is a net agricultural exporter while Taiwan is a net importer of agricultural commodities. From the U.S. perspective, both Mainland China and Taiwan have been major customers of U.S. agricultural exports. In 1989 (fiscal year ending September 30, 1989), Mainland China imported $1,494 million of agricultural goods from the U.S., while the comparable figure for Taiwan was $1,594 million. They ranked, respectively, as the sixth and seventh largest importing countries of U.S. agricultural goods, following Japan, the USSR, Mexico, South Korea, and The Netherlands.

Table 1 shows recent statistics of the values of agricultural exports and imports in Mainland China and Taiwan. Before using the published trade statistics in Mainland China, one has to carefully study the commodity classifications. In several published official statistical books, agricultural trade data include such raw materials as coal, petroleum, leather, and furs. Tuan and Shi (1990) have carefully constructed their trade statistics covering only agricultural products for Mainland China. Their constructed statistics are used in Table 1. For Mainland China, these statistics reflect their agricultural trade levels beginning with the economic reforms of 1979. During this period, Mainland China widely opened its market to exports and imports. During 1983-1989, total agricultural trade (in terms of total value of exports and imports) almost doubled. The growth of agricultural trade was also considerable in Taiwan during this period. However, the patterns of growth are quite different. In Mainland China, the growth has been mostly in agricultural exports, while in Taiwan, the growth in imports has been much more rapid. In 1989, the total value of agricultural trade in Mainland China was about twice the value in Taiwan.

Most of Mainland China's agricultural exports go to Hong Kong, Japan, and the USSR. Even though there are no official statistics on the agricultural trade between Mainland China and Taiwan, Mainland China's agricultural exports to Taiwan have been growing and are expected to expand considerably in the future. These interrelationships of agricultural trade have important implications for U.S. agricultural trade in the Pacific Rim

TABLE 1 Agricultural Trade Balance in Mainland China and Taiwan[a]

	Mainland China			*Taiwan*		
Year	*Exports*	*Imports*	*Total*	*Exports*	*Imports*	*Total*
1983	4.55	3.90	8.45	1.93	3.45	5.38
1984	5.23	2.75	7.98	2.06	3.79	5.85
1985	6.28	2.44	8.72	2.10	3.38	5.48
1986	7.11	2.73	9.84	2.84	3.72	6.56
1987	8.02	3.88	11.90	3.61	4.82	8.43
1988	9.45	5.82	15.27	3.69	5.84	9.53
1989	9.70	6.70	16.40	2.54	5.88	8.42

[a]Unit: U.S. $ billions

Sources: Francis C. Tuan and Shi Ru, "PRC Agricultural Trade Developments and Prospects: With Emphasis on Pacific Rim Countries," paper presented at the International Agricultural Trade Research Consortium, Hawaii, August 1-2, 1990 and Council of Agriculture, *Agricultural Trade Statistics of Republic of China* (Taiwan), 1990.

countries. First, the U.S. has been the major foreign supplier of agricultural products for both Mainland China and Taiwan. In my view, this trend will continue in the near future. However, another important trade relation is emerging. That is, Mainland China and the U.S. are likely to compete in their exports of agricultural goods to Taiwan and Japan. It is true that Mainland China and the U.S. are not exporting the same agricultural commodities. But for products like soybeans and cotton, both Mainland China and the U.S. are the main exporters in the world market. There will be competition for those products. This competition may eventually extend to South Korea when Korea and Mainland China establish closer economic and diplomatic relations in the future. For other agricultural commodities such as pork, Taiwan has indeed already joined Mainland China and the U.S. in the competition in exports to Japan.

International trade theory suggests that a country's trading pattern should be most effectively determined by its comparative advantages in producing goods and services. However, this theory of comparative advantage can explain only partially the trading patterns prevailing in the world market. In many countries, the agricultural sector is highly protected and subsidized. Macroeconomic policies may dictate what agricultural commodities to export and import. Mainland China and Taiwan are no exceptions to this rule. The following two chapters address these and other agricultural trade relationships for Mainland China and Taiwan. A brief summary of these two papers and two commentaries is presented here.

Trade and Macroeconomic Policies

It is interesting to note that both Mainland China and Taiwan have, in recent years, attempted to expand their agricultural trade to achieve macroeconomic policy objectives. However, the underlying economic forces are quite different. Niu Ruofeng contends that Mainland China has been attempting to expand its export of agricultural goods to earn more foreign exchange for its domestic economic development despite the fact that this expansion is made at the expense of domestic consumption. Therefore, Mainland China is adopting a traditional economic development strategy to rely on the agricultural sector to generate capital (in terms of foreign exchange) for its industrialization. The same strategy was also adopted in Taiwan during the 1960s and early 1970s.

In the case of Taiwan, Hsi-huang Chen argues that the huge trade surpluses have generated much external pressure (especially from the U.S.) requesting Taiwan to open its domestic market for foreign agricultural products. As a result, Taiwan has been importing more and more agricultural commodities from abroad (mostly from the U.S.) to reduce its trade surpluses. Along with this macroeconomic policy objective, Taiwan has been trying to upgrade its agricultural structure by discouraging the domestic production of less economical crops such as rice and sugar and encouraging the production of more economical products such as cut flowers. However, this redirection has been very slow and painful. Notably, trade liberalization has also relaxed the government protection of agriculture in Taiwan. In fact, Taiwanese agricultural protection in terms of subsidies for major agricultural commodities has been considerably less than that in South Korea and Japan.

Comparative Advantages

Chen applied and computed several measures of comparative advantage for major agricultural commodities in Taiwan. He shows that Taiwan has been losing its comparative advantage in producing such crops as rice, sugar, soybeans, and corn. There remains only a handful of agricultural commodities with a comparative advantage. These include bananas, citrus, cut flowers, and hogs. Chen argues that Taiwanese farmers do not necessarily conform to such criteria as domestic social cost for social efficiency. He also argues that rice can be profitably produced under government protection.

In the case of the Mainland, Niu's analysis of comparative advantages is not based on such theoretical measures as Chen's is for Taiwan. This is the major criticism voiced by Shin-Hsing Yeh. Based on his judgment alone, Niu argues that Mainland China has comparative advantages in such special products as silk, tea, soybeans, seafood, bamboo shoots, woolen carpets, embroidery, and Chinese medicines. Some of these products appear to make intuitive sense. Nevertheless, products such as soybeans, seafood, and even bamboo products will face increasing domestic demand and perhaps international competition as well. A more careful analysis of Mainland China's comparative advantages would be needed to assess its competitiveness in export trade of agricultural commodities.

Prospects

As pointed out by both Chen and Niu, macroeconomic policies will likely continue to encourage expansion of agricultural exports in Mainland China and agricultural imports in Taiwan. The other overriding policy consideration is food security and self-sufficiency. In his comments, Zhu Daohua argues the importance of self-sufficiency of food for social stability. Zhu voices a strong defense for agricultural protectionism. In reviewing Chen's chapter, he indicates that self-sufficiency in food is a must for Mainland China, even though it may not be for Taiwan.

In this regard, it is important to know the future prospects of Mainland China's grain trade. Currently, the Mainland imports grain in considerable quantity. Niu argues it is more economical to raise grain output (production) than to import grain directly. In fact, future agricultural trade would be affected by likely changes in domestic food consumption. If Taiwan's experience can be a useful indicator, Mainland China will, in the future, decrease its demand for food grain but increase its demand for feed grain. During the last two decades, rising household incomes in Taiwan have considerably reduced consumer demand for rice. In the meantime, heightened demand for meats and dairy products has induced the development of a livestock industry in Taiwan. Consequently, demand for feed grain has increased dramatically. This has been the force underlying the sharp increases in Taiwan's grain imports in recent years. If Mainland China follows this path, one may expect continued growth in Mainland China's grain imports in the future, a conclusion supported by Dwight Perkins' analysis (Chapter 10).

8

China's Agricultural Trade and Prospects in Foreign Competition

Niu Ruofeng[1]

Agricultural imports and exports comprise an important part of Mainland China's foreign trade. At present, China's relatively poor technologies hamper the ability of the industrial sector to produce and export large quantities of manufactured goods. Therefore, the promotion of agricultural trade and development of export-oriented agriculture are not only important objectives of China's agricultural modernization, but also ways to earn more foreign exchange needed to import advanced technologies, equipment, and industrial materials for sustained economic development. Under increasing competition in world agricultural market, China should adjust its trade strategies in accordance with its comparative advantages. This will be a key issue in formulating China's agricultural trade policies.

An Overview

In the early stages after the founding of the People's Republic, the volume of China's agricultural trade was minute. It grew vigorously thereafter, particularly during the past decade. The total export value of raw and processed agricultural and sideline products exceeded $19 billion (in U.S. dollars) in 1988. Generally speaking, Mainland China's agricultural import and export trade is still relatively underdeveloped. Its total agricultural trade volume is only one-fourth that of the United States. Only recently, China became one of the ten largest countries in terms of agricultural trade volume. However, China has occupied an important place in the world trade of selected agricultural products. For example, in 1987, China was the largest exporter of silk, fresh eggs, and peanut kernel; and the second largest exporter of live pigs, while the exports of cotton, tea, and soybeans were also among the largest in the world. While exporting rice, soybeans, and pulses, China imports certain kinds of cereals in large volume. China is the second largest importer of wheat following the Soviet Union, and is among the biggest buyers of sugar and wool in the world market.

China's agricultural trade has been stimulated by long-term fluctuations in domestic agricultural output and a tight balance between domestic production and consumption of many farm products. The following features of China's agricultural trade are distinctive from those in the developed countries with a large surplus volume of agricultural products.

1. *Agricultural exports and share of total export value.* Except during the early 1960s and a few other years, the agricultural exports have experienced an upward trend during the past four decades (Table 8.1). The exports of raw and processed agricultural and sideline products were valued at $16.3 billion in 1987, 18.6 times and 1.7 times over the levels of 1953 and 1978, respectively. Although the agricultural trade plays an important role in the nation's foreign trade and ranks third in terms of the foreign currency it earns, the proportion of export value of raw and processed agricultural and sideline products in total export earning has been declining over time. During the 1950s and 1960s, the export value of raw and processed agricultural and sideline products accounted for about three-quarters of the total export value. This share decreased to 60 percent in the mid-1970s and to less than 50 percent after 1985. These changes in Mainland China's trade structure mainly resulted from the rapid increases in the exports of manufactured goods (Table 8.2).
2. *Exports vs. domestic consumption.* Currently, per capita consumption of major food items in China is well below world average levels (Table 8.3). However, to earn more foreign exchanges, China still attempts to expand exports of some staple agricultural products that have been in short supply domestically. For example, although the per capita consumption of meat is less than 20 kg, or just two-thirds of the world average level, China exported 3 million heads of live pigs and 100 thousand tons of frozen pork in a recent year. A similar situation exists for cotton. China exported 560 thousand tons and 760 thousand tons of cotton in 1986 and 1987, respectively. Some high-valued goods such as sea cucumber are given priority to export. Thus, for some commodities, a decline in domestic production would create a severe shortage in domestic market.
3. *Export of low value farm products.* In 1987, the ratio of export values of raw and processed agricultural and sideline products was one to two. Due to the outdated technologies of processing, preservation,

TABLE 8.1 Imports and Exports of Major Farm Products in Mainland China (1,000 tons).

Year	*1952*	*1957*	*1965*	*1978*	*1980*	*1984*	*1985*	*1986*	*1987*	*1988*
Exports										
Frozen pork	29	54	143	43	64	100	111	105	100	63.5
Live hogs (1000 heads)	779	507	1,719	2,463	2,468	3,080	2,960	3,100	3,020	3,030
Aquatic products	23	58	120	92	106	110	128	167	218	287
Tea	29	41	38	87	108	140	141	172	174	198
Soybeans	865	1,141	653	113	114	840	1,140	1,714	1,301	1,480
Cotton	6	34	35	34	10	2	3	558	755	468
Filature silk	2	3	3	9	8	9	11	9	9	9.4
Imports										
Grain	a	167	6,405	8,883	13,429	10,445	6,170	7,730	16,280	15,330
Cotton	77	48	199	509	898	35	a	a	6	34.8
Edible fats & oils	1	35	81	330	355	120	200	470	850	793
Wool	b	b	b	b	b	56	113	152	153	187.4
Sugar	53	119	208	1,299	912	1,228	1,870	1,180	1,830	3,710

[a] Quantities are less than 0.5.
[b] Data not available.

Sources: State Statistical Bureau, *China Statistical Yearbook*, China Publications Company, 1986, 1988, 1989; and State Statistical Bureau, *China Statistics of Trade and Prices*, 1952-1983.

TABLE 8.2 Changes in Mainland China's Trade Structure Between 1952 and 1988

Product	1952		1988	
	Import Share (%)	*Export Share (%)*	*Import Share (%)*	*Export Share (%)*
Primary Products (including raw and primarily processed products)	18.7	83.4	18.2	34.4
Finished Products	81.3	16.6	81.8	69.6

Source: Economic Daily, September 26, 1989.

TABLE 8.3 Per Capita Consumption of Mainland China's Major Items of Food and Clothing

Products	*Unit*	*1978*	*1980*	*1985*	*1986*	*1987*
Grain[a]	kg	195.5	213.8	253.2	254.8	251.4
Edible Oil	kg	1.6	2.3	5.1	5.2	5.4
Pork	kg	7.7	11.2	13.9	14.3	14.5
Fresh Eggs	kg	2.0	2.3	5.0	5.2	5.6
Sugar	kg	3.4	3.8	5.6	6.1	6.7
Clothing[b]	Meter	8.0	10.0	11.6	11.1	11.3

[a]The figure of grain refers to processed weight.
[b]Clothing includes pure cotton cloth, cotton blends, and pure synthetic fabrics.

Source: China Statistical Abstract, 1988.

packaging, and storage, the qualities of both raw and processed products exported are relatively low, and so are the prices. For instance, Chinese rice was sold on the world market at $312/ton in 1987, which was 37.4 percent less than the rice from Thailand. China has been exporting a large proportion of raw and unprocessed farm products at low prices.

4. *Fluctuations in agricultural trade.* The fluctuations in China's trade volume reflect variations in both domestic agricultural output and national economic and agricultural policies. These relationships can be best illustrated by the case of grain trade during the period 1949-87. China was a net grain exporter during the 1950s, when the grain production was first recovering from the damage of the civil war and then reaching a stage of rapid growth in production. She owed this achievement to national policies for coordinated development of different branches in the economy. In 1957, total and per capita grain output rose to 195 million tons and 309 kg, respectively, representing an increase of 72 percent and 48 percent, respectively, over 1949 levels.[2] Although income and living standards were also low during this period, China was able to export grains to earn foreign exchange urgently needed for the national industrialization program. Net grain exports showed an upward trend during that period, reaching to 4.2 million tons in 1957.

Domestic grain output plummeted in the early 1960s, largely owing to the fault of such policies as the "great leap forward" and the establishment of the people's commune. Per capita grain output dropped to 238 kg in 1962, 23 percent lower than the 1957 level. Although grain production gradually climbed during the three-years economic readjustment (1963-65) and grew slowly thereafter, rapid population growth brought per capita grain availability to a virtual standstill, fluctuating around the 300 kg level. At the same time, the per capita production of other farm products was also lower than the 1951 level. From 1961, China became a net grain importer, with annual imports ranging from a low of 600 thousand tons to a high of 5.7 million tons during the period of 1961-77.

Since the start of the economic reforms in 1978, the government adopted a series of policies to promote the import of about 40 million tons of grain to rehabilitate China's ruined rural economy and to assist the development of livestock raising and food processing in rural areas. The rural reform and other favorable factors have resulted in a boom of the rural economy in China. Grain output registered a phenomenal increase of 33.6 percent in the period 1978-84, or specifically from 304.8 million tons in 1978 to 407.3 million tons in 1984. Meanwhile, the per capita grain production reached 396 kg (in raw weight) in 1984, resulting in a partial relaxation of the long-standing tight balance between domestic grain production and consumption.

Although the upsurge of grain production in the early 1980s turned Mainland China from a net importer to a net exporter of grain in the short period of 1985-86, an excessively optimistic judgment on the potential of further agricultural growth resulted in several unreasonable and unrealistic changes in agricultural policy. Specifically, the policy changes attempted to readjust the structure of planted areas between staple and nonstaple food crops. As a result, the planted area for staple food crops was reduced by 1.5 million ha in 1985. These policy errors caused the stagnation of grain production in the successive four years following 1985 and made China again a large importer of grain.

The preceding analysis reveals that the fluctuations in China's grain trade have been closely related to domestic grain production and policy changes. China has, for a long time, carried out a heavy industry-oriented development strategy. Under such a biased strategy, agricultural production and trade have been in a subordinate position to the needs of development in heavy industry. Moreover, there have been fluctuations in agricultural output. Therefore, China's agricultural trade has been unsteady during the last four decades.

Strategies and Measures

What will be the appropriate agricultural trade strategies and policy measures for Mainland China? To answer this question, we need to consider factors affecting the balance between domestic demand and supply, foreign exchange payments, and changing trends in the world market.

In my view, the shortages of agricultural supplies in China will persist in the long run. It seems reasonable that China will continue to face great pressure for increasing consumption owing to the large and growing population, limited per capita land and other input resources, relatively low agricultural productivity, and limited supply expansion ability. According to the projections made by the Chinese Academy of Agricultural Sciences (CAAS), by the year 2000, the Mainland Chinese population will exceed 1.25 billion and the per capita income will be doubled (the per capita annual disposable income was 119 *yuan* in rural areas and 545 *yuan* in urban areas in 1988), causing upward pressure on the demand side. The report also indicated that when people spend half of their disposable income on food, the per capita annual grain demand will reach at least 425 kg, implying a total demand for grain of at least 531 million tons a year. Similarly, the aggregate demand for cotton is estimated to rise to 5.7 million tons by the year 2000. To meet these targets, the output of grain and cotton should grow at a rate of at least 2 percent annually from 1988 levels. Considering the diminishing trend of cultivated land area and increasing scarcity of other agricultural resources, the balance between demand and supply can hardly be realized. For this reason, gaps between domestic production and consumption are expected to persist in the future.

Second, the primary objective of the agricultural development strategy and a long-term government policy should be to ensure stable and sustained growth of agricultural production, and thus, to increase the effective supply of farm produce. The precondition to guarantee this growth is to have stable government policies in favor of the agricultural sector. Therefore, the policy of heavy-industry-oriented development should be adjusted and replaced with a strategy of harmonious development of both agriculture and industry. In fact, both industrialization and agricultural modernization need to proceed simultaneously in China. To develop a modern agricultural sector and to ensure the people a good living standard by the end of this century would increase the strength of the nation. In view of the limited reserve of foreign exchanges and surging food demand, it is impossible for China to import a great quantity of grain. In fact, grain import could only serve as a supplemental temporary solution to meet increasing food demand. Any opinion that China can solve its food problem by importing large quantities of foodstuffs in the world market seems illusory at best.

Third, to ensure adequate supply of exportable commodities and to improve their quality, China should make every effort to build up the production bases of exportable farm products, such as soybeans, tea, fruits, aquatic products, silk, lean-meat pigs, rabbits, and other specialty products. It is widely believed in China that the development of these production bases is conducive to solving such problems as undue dispersion, poor quality, and lack of standardization, and unrealized potential for specialization and scale economies. The administration of Production Bases for Export Commodities under the Ministry of Foreign Economic Relations and Trade is responsible for coordination between the trading companies and government planning and administrative agencies in making production and sale contracts down to the county level. Since 1986, the total amount of state and local government investments has reached 1.24 billion *yuan*, with 576 agricultural commodity bases being established. In 1988, these joint investments amounted to 0.56 billion *yuan*, funding the projects to build up the foundation for another 237 agricultural commodity bases. These bases are spread among many crops, including 191 grain bases, 74 high-quality cotton bases, 154 bases producing high-grade edible oil, fruits, vegetables, famous teas, mushroom, flowers, and traditional Chinese medicine, and

394 bases for rare and specialty products. Many of these development projects have been highly successful. For example, the total output from 60 high-quality fruit bases located in Liaoning, Hebei, Shandong, and Sichuan provinces, has increased by 45 percent. In the Wafangdian area in Dalian, Liaoning province, where apple production ranks first in the country, the production base developed an apple variety producing at least 80 percent of high-quality (first-grade and second-grade) fruits. Consequently, the ratio of good quality apples has reached 99.2 percent. These high-quality products provide excellent sources for export earnings.

Meanwhile, the reform of the foreign trade administrative system has made great progress in recent years. Some production bases in coastal provinces like Jiangsu, Guangdong, and Fujian have abolished the old management system characterized by separate farm production, processing and trade and substituted an integrated system that all partners at different stages of agricultural production and marketing share the same benefits and risk. This new management system is likely to enhance China's competitiveness in the world market.

Fourth, there is an important linkage between the adjustment of production mix and improvement of product quality. As noted earlier, the previous strategy of exporting rawfarm products at relatively low prices had long put China in a unfavorable position in the competitive world market. The structure of exportations should be adjusted in the direction of exporting more manufactured goods instead of raw materials, and more industrial commodities and less farm produce. It is predicted that the rank of agricultural exports in total export value will shift from the third largest to fifth, following light industry, the textile industry, electrical machinery, and crude oil and coal. As for the export structure, our guiding precept is to reduce the exports of low-grade and low-value products, while increasing exports of high-grade and high-value products. There are two key tasks for increasing agricultural exports, namely to develop improved varieties and to upgrade technologies for processing, packaging, and preservation. If we fail in these tasks, China will face a more unfavorable environment of international competition.

Fifth, China's comparative advantages and great potentialities in agricultural trade lie in her endowment of plentiful natural resources and specialized input factors, including an abundant labor supply and many unique processing skills. Some special products, such as silk, famous brands of tea, soybean, seafood, bamboo shoots, bamboo products, handmade woolen carpets, embroidery, patent medicines, and handicrafts are in great demand in the world market. China should make efforts to improve the quality and renew the varieties of such products in accordance with the changing tastes and preferences of foreign buyers so as to expand overseas markets.

Finally, Mainland China should further diversify its trade relations. To date, China has established and developed trade relations with more than 180 countries and regions. Moreover, cooperative labor transaction relations exist with more than 110 countries and regions. During recent years, our trade with various countries and regions has increased in all dimensions. According to the 1987-1988 statistics published by the Editorial Board of the Almanac of China's Foreign Economic Relations and Trade (1989), among China's largest trade partners were Hong Kong, Japan, the European Economic Community (EEC), and the United States. The volume of barter trade with the Soviet Union and East European countries was valued about $5.6 billion in an approximate balance of trade. The trade with the Middle East, Latin America, Africa and Southeast Asian countries has been expanded gradually in recent years. The frontier trade with the Soviet Union and other neighboring countries has also witnessed a vigorous development. There are great potentialities in the reciprocal trade between China and the Soviet Union; total trade value amounted to SF 4 billions (Swiss Francs) during 1986-1988. The frontier trade in 1988 grew so rapidly that the total trade volume was 3.1 times of that in 1987, and the volume of trade on the western frontier alone amounted to $653 millions (U.S. dollars). I believe that the current economic adjustment, steady growth of demand in the western economies, and the normalization of relations with the Soviet Union will provide China favorable opportunities to expand her foreign trade. In addition to the old partners, China should also develop commercial relations with the Soviet Union, East European countries and other Third World countries. Moreover, Mainland China is active in direct reciprocal trade with Taiwan. It has been reported that the value of such trade exceeded $1.1 billion in 1987 and topped $1.74 billion at the end of the first half of 1989, representing an increase of 57 percent over that of the preceding year. I am deeply convinced that the reciprocal trade between Mainland China and Taiwan presents vast vistas.

Prospects

In my view, the strategy to modernize China's socialist economy can be divided into three steps. The first step would be to double GNP from the 1980 level to meet increasing food demand. This step was achieved in 1989. The second step is to double GNP again by the end of this century to provide a comfortable living standard for Mainland citizens. The third step is to achieve a full modernization during the next century. By then, the per capita GNP should reach the level of average medium-developed countries. During this century, with a steady expansion of the agricultural sector,

Mainland China's agricultural trade should enjoy a tangible development. According to the projections by the Editorial Board of the Series of China, in the year 2000, the total export value of agricultural and sideline products (excluding processed goods) will reach $9.5-10 billion (in U.S. dollars) with an annual growth rate of 5.5 percent, and will make up about 10 percent of total export value. If this projection holds, it would mean that agricultural exports should increase by $2.2-2.4 billion from the 1987 base level ($5.88 billion). This is a possible target if the government has stable national policies favorable to the agricultural sector.

Among staple agricultural products, the superiority of exporting grain of improved variety may diminish or even disappear. A better prospect is to process high-grade products and unique traditional products. For example, recently processed chicken has had a high demand in the world market. Cultivated shrimp is popular in Japan, Canada, the United States, and Europe. In addition, exports of tinned food, peanuts, and fruits to Hong Kong, the Soviet Union, and Eastern European countries have great potential. I believe the Mainland should increase the exports of fruits, vegetables, flowers, and plants to neighboring countries. Qing Dao beer, Five-Star beer, and Long Xiang beer are good export items to be promoted in the world market. Meanwhile, traditional products such as famous teas, Chinese medicine, filature silk, eiderdown, and rabbit hair should maintain their shares in export markets. Other famous local products and handicrafts have established buyers in Southeast Asia and some other regions strongly influenced by Chinese culture, especially where many overseas Chinese reside.

China's textile industry possesses large processing capacity, and domestic demand for cotton is strong. It is expected that China will not export much raw cotton in the future. Instead, the export of natural silk and fabric will increase.

Grain and oil-bearing crops should retain a supplementary nature for the purpose of regulating varieties and filling the gap of domestic demand and supply. Chinese rice is relatively less competitive in the world market. The only reason to export rice is to exchange it for wheat. In 1988, rice exports reduced sharply. Consequently, the Mainland became a net rice importer. The shortage of feed in the domestic market may restrict exports of corn and other coarse grains that contain high levels of protein but low foreign exchange values. Meanwhile, imports of wheat and other food grains may rise slightly from the 1987-88 level. However, as there is hardly any difference between the prices of chemical fertilizer and wheat, it is more economical to import urea to raise grain output than to import grain directly. Oil crops such as rapeseed, peanut, and sesame are in such ample supply that it is necessary to reassess the need to import several hundred thousand tons of edible fat and oil annually to meet the domestic demand.

With good natural conditions, Hainan province can expand its production of, and even export palm oil. The shortage of sugar and milk products will remain in the short run, so we should take advantage of the ample supply and lower price of sugar in the world market by increasing imports and developing more profitable crops with our limited arable land. In the meantime, China should also take the opportunity in this era of overproduction of milk in the world market to import more milk powder, to be reconstituted to meet domestic demand. By doing this, the Mainland could use more forage for the production and export of beef.

After 10 years of economic and agricultural reforms and an open-door policy, Mainland China's foreign trade has been booming in all directions and at all levels. It can be predicted that agricultural trade between Mainland China and her major trade partners will continue to expand. However, compared with other countries in the world, the agricultural sector in China will retain its strongly self-sufficient character for quite a long period. A major challenge to the Chinese government is to ensure an optimal amount of grain imports, and to set bounds on the export of high-grade agricultural products.

In conclusion, China may hardly become a large exporter of staple agricultural products. Admittedly, there are some unfavorable obstacles in the development of foreign trade, such as output instability, poor quality, and low transport capacity. Moreover, China's ability to gain foreign exchange has declined because of the increase in domestic purchasing prices which, in turn, has raised the prices of export products.

In view of the above, a series of comprehensive and effective steps must be adopted to remove these obstacles. Further efforts should be made to establish commodity bases for foreign trade and to improve product quality. The government should also take measures to draft proper laws and regulations to ensure and protect trading contracts. More attention should be paid to improving capacity and the technologies for preserving, processing, and transporting agricultural products. Only in these ways can China enjoy higher international prestige in the world market and make greater contributions to the development of international trade.

Notes

1. The author would like to admit that agricultural trade is not his major research field. The purposes of his chapter are to show the current situation of China's agricultural trade and to share personal opinions on China's trade policy issues.

2. In Chinese statistics, grain output usually includes all cereals, soybean, potatoes, and sweet potatoes (converted to grain equivalents on a 5:1 basis) at rough weight. However, as far as the import of grain is concerned, only cereals are included in accordance with FAO statistics.

References

Editorial Board of the Almanac of China's Foreign Economic Relations and Trade, 1989, *Almanac of China's Foreign Economic Relations and Trade 1986-1988*.
Editorial Board of the Series of China in the Year 2000, *China's Economy in the Year 2000*, forthcoming.
State Statistical Bureau, 1989, *China Statistical Yearbook 1988*, China Publication Company.

Discussion

Shin-Hsing Yeh

It is my great pleasure to participate in the symposium on Chinese Rural Development, organized by the AAEA, and to present my brief comments on Professor Niu Ruofeng's chapter titled "China's Agricultural Trade and Prospect in Foreign Competition."

Professor Niu's chapter has five sections. The first describes the importance of agricultural imports and exports in China's foreign trade. Section Two is an overview of China's agricultural trade. It presents a detailed description of the past and present circumstances of agricultural trade in Mainland China. Section Three formulates agricultural trade strategies and measures. The design of these strategies and measures is based on particular conditions in China. Section Four attempts to forecast the future development of China's farm product exports. The last section contains concluding statements.

Generally speaking, Professor Niu's analysis is straightforward, brief, and to the point. However, there are areas of confusion regarding both content and assertions that I address below.

First, Professor Niu points out in his introduction that under increasing competition in world agricultural trade, Mainland China should adjust its trade strategies in accordance with its comparative advantages. However, as we know, comparative advantages or disadvantages may change over time. Since Professor Niu did not employ any method to examine the changes in comparative advantages, his view on the relative advantages of various agricultural commodities has to be considered as his personal, subjective judgment. The estimates of the changes in comparative advantage would have provided important information about the changing patterns of agricultural structure and production (such as labor costs, product characteristics, and quality). In addition, the paper does not analyze the changes in comparative advantages in various international markets or the changes in its competitiveness with other countries.

I strongly suggest that the revealed comparative advantage (RCA) index be used to test the comparative advantages of farm products. Furthermore, the time-series data of the RCA index could compute the coefficient of variation, a useful measure of the dynamics of change. To compare the competitiveness, in various foreign markets, of farm products from Mainland China, the constant market share (CMS) method could be used.

Second, the author asserts that the present relative poor state of technology in China is hindering the industrial sector's ability to export sufficient quantities of manufactured goods. The author also asserts that to earn more foreign exchange for national economic development, China needs to promote agricultural trade and to develop an export-oriented agricultural sector. It seems to me that this reasoning does not hold because of the following facts:

1. The scale of management in the operation of Chinese farms is comparatively small and production costs are relatively high. Moreover, the skills used in the production and processing of agricultural products are backward, and the distribution network is not well advanced. Hence, China's farm products seem to have difficulty in gaining comparative advantage in international markets except in the case of some traditional items.
2. Farm products have less of a comparative advantage than industrial products in terms of trade in the international market. Therefore, if China depends highly on agricultural trade to earn more foreign exchange to import advanced techniques, equipment, and industrial materials, it would have difficulty in achieving this objective.
3. The price elasticity of demand and the income elasticity of demand of farm products are perhaps much lower than those of industrial products in the importing countries. Therefore, it would be difficult to increase foreign exchange earnings through efficient export of farm products.

Third, because the terms of trade of farm products are inferior to manufactured products in the international market, China has to export more and larger amounts of farm products, and perhaps lower their prices, to earn more foreign exchange. To achieve this goal, China may have to lower its domestic procurement prices of farm products, which will deprive farmers of profit in the production of farm products. Besides, a considerable expansion in the export of farm products would lead to a serious shortage of farm products in domestic markets. This, in turn, may accelerate inflation in China's domestic market, and this high inflation could adversely affect the stability of the economy. Based on the above reasons, China faces a dilemma in its exports of domestically needed farm products, both now and in the future, to procure foreign exchange; such exports may end in inflation.

9

Taiwan's Agricultural Trade and Prospects in Foreign Competition

Hsi-huang Chen

Introduction

Since World War II, Taiwan has been quite successful in agricultural development. Growth in agricultural output was achieved chiefly through technological innovations that raised the productivity of the relatively fixed inputs of land. Labor and capital, in combination with land, are two major factors of agricultural production that are substitutable in the process of agricultural development. In the early stages of development, redundant labor served as a cheap source of growth in Taiwan. Farming was usually performed by labor-using and land-saving technologies. At that time, compulsory methods were used to accumulate capital, often at the expense of farmers' living standard. However, as the nation pursued its development, profitable new technologies were gradually introduced, and government policies on prices, inputs, and credit provided solid economic incentives to farmers. This combination of factors has resulted in increasing income for some farmers and the possibility of relying more heavily on voluntary means for capital involvement in substitution of labor. Thus, farming methods have successively changed from labor-using and land-saving to capital-using and land-saving.

As Taiwanese agriculture develops, farm and nonfarm commodity markets become increasingly interdependent. Farmers produce agricultural products not only for their own consumption but mostly for market sales. Because of a close relationship between agricultural production and food processing, the more the agricultural sector advances, the greater its dependence upon the development and modernization of the processing industry. More important, a well-developed food industry promotes not only the exports of processed foods but also the imports of raw agricultural products. Such "import substitutions" and "export promotions" are induced on the basis of the comparative advantages of the countries involved.

In recent years, due to the changes in world and domestic economic situations, Taiwan's economy has encountered tremendous pressure to comply with economic deregulation and trade liberalization. Because of the huge trade surpluses, despite a surging value of New Taiwan Dollars (NT$), the economy's current-account balance soared to more than U.S.$ 70 billion in 1988. Thus, Taiwan is recognized as rich by the international economic community. Taiwan is therefore asked to open up its domestic market for foreign products and to pursue the economic liberalization or deregulation as a means to share the burden of balancing the world market. In facing these new restraints, agricultural production policies have to meet challenges and pressures from abroad, especially the United States.

Taiwan is a densely populated country, and agricultural production is particularly important to national security. However, with the agricultural sector growing weak, agricultural products have difficulty competing with foreign farm products in both the domestic and world markets. Particularly, the increasing value of the NT$ has widened the gap between domestic farm prices and foreign farm prices. Such exchange rate increases have reduced the competitiveness of domestic agricultural production. It is anticipated that the agricultural sector will shrink further if no new adjustment strategy is adopted.

This chapter examines the effects of opening up domestic markets and strengthing the NT dollar upon Taiwan's agricultural production and trade, and discusses the possible policies that should be adopted in responding to this new economic environment.

Agricultural Policy and Trade

After World War II, Taiwan's agricultural policy has pursued two objectives—food security and farmers' welfare. To fulfill these policy objectives, the government has emphasized the necessity of reducing production costs and boosting farmers' income. The general agricultural policy adopted by the government has centered on its chief goals of maintaining low prices on farm requisites and guaranteeing reasonable prices for farm products. Under this policy, farmers are given a real incentive to increase crop production.

With respect to production policy, the government has paid more attention to food crops than other agricultural products. Rice, as the main staple food for Chinese people, is produced with a view to achieving self-sufficiency. Corn and soybeans are mainly used for feeding livestock and poultry, and they are considered as two important agricultural commodities on Taiwan's overseas shopping list. Sugar and bananas, two long-time export items of Taiwan, earned much in foreign exchange before 1980, but have lost their important position among exports in recent years.

The government's rice policy aims at satisfying the demand for rice fully with domestic production. This would require that optimum incentives and support be provided so that the growth of rice yield could always surpass population growth. With respect to corn and soybeans, the government does not have any intention of meeting the demand with domestic production. However, farmers are still encouraged to produce corn and soybeans to (1) improve the land utilization rate, and (2) reduce dependence upon foreign supplies of feed to the domestic livestock and poultry industries. At present, Taiwan still has some idle land that can be exploited for growing more corn and soybeans. However, as the CIF prices of imported corn and soybeans are lower than production costs of these two crops, market prices do not constitute any incentive for farmers to increase corn and soybean production.

For sugar and bananas, the government has been providing incentives for production based on prices in the world market. Sugar production and marketing are handled by the Taiwan Sugar Corporation (TSC). While the domestic sugar market is open to sugar cane farmers and sugar dealers in free competition, the TSC may intervene in the domestic sugar market to protect the interests of both producers and consumers by adjusting the quantity of sugar for trade in the local market. Generally, the TSC attempts to maintain the domestic sugar price at a relatively high level as an incentive to encourage farmers to plant more sugar canes.

Government provision of incentives for banana production are generally not effective. Generally speaking, an increase in the price received from export would stimulate both banana production and exports, because it is rather easy for farmers to expand banana acreage and to export bananas originally reserved for domestic consumption. But the actual situation is not so, because the government has enforced the retirement of banana plantings on steep slopes for the sake of soil conservation; and furthermore, bananas are currently exported only to Japan, which maintains a strict banana import quota. The problem in banana exports is not price, but market limitations. As efforts have been made by the government to reduce banana acreage and production in recent years, it is unlikely that an increase in foreign market price would increase the total production. It is possible that an increase in price in the foreign market would drive up the price of bananas in the domestic market and exporter profits, but it can hardly affect the elasticity of banana exports for reasons previously stated.

In Taiwan, agricultural trade policy is always linked to national agricultural policy. With respect to food imports, cereals (mostly wheat and corn), oilseeds (chiefly soybeans), dairy products, and manufactured tobacco account for the bulk of Taiwan's agricultural imports. Of total agricultural imports, during the period 1977–1987, the food component increased from U.S.$ 852.2 million to U.S.$ 3,416.7 million, mainly due to a growing animal industry, which in turn brought on a higher demand for imported cereals. Most cereal processing industries in Taiwan depend mainly upon imported raw materials. Specifically, manufactured foods (mostly for domestic consumption) such as soybean products, flour, and edible oil, are based on raw materials, of which more than 90 percent are imported. In comparison, exported manufactured foods depend mostly upon raw materials domestically produced by farmers.

As for canned foods, all food canneries qualified for export are registered with the government's foreign trade authorities as exporters. To ensure an orderly marketing of certain canned foods such as pineapples and mushrooms, an export quota for each cannery is set by the foreign trade authorities. However, food canneries are free to process and export other canned foods. Most of them handle the export business themselves, but a number of small food canneries sign contracts with big food canneries to handle the export business on their behalf. Some food canneries also sign contracts with well-known foreign firms in importing countries or in Taiwan to be their selling agents to expand the foreign market.

Taiwan, as an economy with scarce natural resources, relies heavily upon international trade to help promote its agricultural and overall economic development. The agricultural trade of Taiwan was valued at U.S. $ 8.4 billion in 1987, constituting 9.6 percent of its total foreign trade volume. Within agricultural trade, imports amounted to U.S.$ 4.8 billion, representing 14.0 percent of total imports, and exports totaled U.S.$ 3.6 billion, or 6.8 percent of total exports.

Since the 1980s, the rapidly growing interdependence in the world market has increasingly integrated Taiwan's agriculture with the world economy. In recognizing this necessity as well as the willingness to share mutual benefits with its trading partners, the government has taken several actions to liberalize agricultural trade activities, such as lowering tariff rates and removing trade restrictions or barriers on the importation of foreign agricultural products. For instance, in 1987 alone, tariff schedules imposed on many products were reduced twice, first in April and again in June, in response to U.S. requests.

Moreover, the uniform import price plan, covering major bulk commodities such as wheat, corn, and soybeans, has effectively insulated importers from international price fluctuations. The policy has also had the impact of lowering the elasticity of Taiwan's import demand for these commodities with respect to changes in world price levels. While the practices of mandated group purchase have helped those inexperienced individual importers in dealing with complicated international trade and given the government some control over bulk commodity imports, the bureaucratic rigidity of the practices has reduced the efficiency of this import operation. The government recently has also substantially relaxed the import quotas on bulk commodities and announced full liberalization of these imports.

In the further pursuit of trade liberalization, tariff cuts involving 3,575 products were made effective in March 1988. Among those, 549 items were farm products. As a result of the cuts, the nominal tariff rate for farm products has dropped from 20.4 percent to 16.8 percent.

Almost all countries protect their domestic agricultural products against imported ones. Comparatively speaking, Taiwan has a quite liberal trade policy regarding farm imports. For instance, Taiwan ranked third among U.S. overseas markets in 1986, and was the number one market for American apples. In 1988, Taiwan produced 1.9 million tons of rice but imported 7.5 million tons of grains. A recent study by the U.S. Department of Agriculture ("Government Intervention in Agriculture") shows that countries like Japan, the European community, the U.S., and South Korea offer their farmers much more protection than Taiwan.

Comparative Advantages of Major Tradable Crops

In comparative advantage analysis, economists concerned with future resource allocation have recently attempted to measure relative trade advantage in terms of foreign exchange earnings. Pearson, *et al.* constructed three measures for this purpose. They are net social profitability (NSP), domestic resource costs (DRC) of foreign exchange earned or saved, and the effective rate of protection (ERP) on value added. All three measures are theoretically sound but differ in analytical concept and assumptions. The NSP contains the adjustment of economic rent and shadow price, and external effects on the local economy. The DRC measures the social cost of earning a net marginal unit of foreign exchange. The ERP measures the percentage increase, resulting from protection, in the domestic value added over the value added in a free trade situation. As noted by Pearson, *et al.*, NSP, DRC, and ERP have been applied or suggested for application in two broad areas — the *ex post* measurement of the cost of trade distorting policies and the *ex ante* evaluation of resource allocation. The methods for computing NSP, DRC, and ERP are detailed in the appendix.

By definition, if trade is socially profitable, then its DRC coefficient should be less than one, and NSP should show a positive advantage; when it is not socially profitable, its DRC coefficient should be greater than one, and NSP should show a negative advantage.

Eight crops—corn, soybeans, bananas, sugarcane, rice, citrus (ponkans), cut flowers (chrysanthemums), and hogs — are selected under this study for a comparative advantage analysis. Corn and soybeans account for the bulk of Taiwan's agricultural imports; bananas and sugar were the main agricultural export products. Rice is the main food crop produced for achieving self-sufficiency but still with a high level of commercialization in the world market. Citrus and cut flowers are two cash crops with increasing potential for exports, and hogs serve as the main food livestock produced not only for domestic consumption but also for export.

For evaluating the comparative advantages of the eight selected agricultural commodities, NSP, DRC, ERP, and related parameters are estimated in two-year intervals from 1970 to 1986. The results are presented in Table 9.1.

Corn and Soybeans

As mentioned previously, the government in Taiwan has no intention of meeting the domestic demand for corn and soybeans with local production. Any policy recommendations as to how much should be produced locally, and with what impact on production, should take into consideration efficiency criteria and the goal of reduced dependency on foreign sources of corn and soybean supplies. These conflicting goals cannot be considered entirely from the economic point of view. Indicators of comparative advantage may be of reference value only but cannot serve as the only factors in policy decision making.

Table 9.1 reveals that the comparative advantages in corn and soybean production, as measured by the DRC coefficient, were good for corn in 1970 and good for soybeans only in 1970, 1972, and 1974. The other years covered in this study witnessed only comparative disadvantages for the production of both corn and soybeans. As can be seen from the time series data, the comparative disadvantages for corn and soybeans have worsened in recent years. However, the supply response of the grain farmer does not necessarily conform to criteria for measuring social efficiency with the DRC coefficients. With government-guaranteed prices, private profitability has positive advantages for corn and soybean production every year. As shown in ERP coefficients, almost all coefficients have the positive protection rate except 1974 and 1976 for soybeans, which explains government protection for domestic production in the form of a price supporting policy. The government policy of guaranteed prices and protection may do harm to resource allocation but will give farmers an additional incentive to continue producing corn and soybeans to achieve the policy target of reducing dependence upon foreign supplies.

Sugar and Bananas

Sugar and bananas are both traditional agricultural export items in Taiwan. Before 1976, sugar had low production costs, and the DRC coefficients were below 1.0, implying a comparative advantage. However, due to the decreasing trend in world market price, sugar production has had a comparative disadvantage, as the computed DRC coefficients have been greater than one since 1976. With respect to bananas, all DRC coefficients are below one (ranging from 0.610 to 0.976), indicating that bananas had a comparative advantage in production for all years under study.

Sugar is subject to relatively high negative effective rates of protection, primarily because of the government's contract to purchase raw materials from cane farmers at a low price to reduce sugar production costs. This contractual arrangement has greatly boosted Taiwan's competitive position in the world sugar market. Taiwan also expanded banana acreage and yield to curtail and reduce the production of other main food crops. Although in the world market the prices of sugar and bananas have fluctuated frequently, farmers are always hoping to see high prices in the world market. Sometimes farmers benefit immensely from very high prices in the world market, with resulting increases in contract price.

In recent years, however, sugar and banana exports have been subjected to strict quotas imposed by importing countries. This trade barrier has generated serious problems. Production must be geared to world market demand. Too much production usually results in lowered domestic prices. Thus, high world market prices may not give farmers a comparative advantage for the production of these two crops because of the quotas. As we can see from Table 9.1, the effective protection coefficients (EPC) are less than one in most years. These figures translate to negative protection as measured by ERP during these years.

Citrus is also subject to relatively high negative effective rates of protection due to the trade liberalization of lowering tariffs for citrus imports.

Rice

For rice, all calculations of trade indicators are based on the Ponlai variety, the most popular in Taiwan. The DRC coefficients for rice were less than one before 1980, but greater than one after 1980. Generally speaking, with the keen competition in today's world rice market, rice production would not be so desirable for resource use. As Taiwan's economic development advances, the DRC coefficients can be expected to grow. However, the DRC coefficient was 0.355 in 1974, the lowest

TABLE 9.1 Indicators of Comparative Advantage[a]

Commodities	*Year*	*NSP*[b]	*Comparative Advantage or Disadvantage*[c]	*DRC*[b]	*Level of Comparative Advantage or Disadvantage*[d] *(1-DRC)*	*EPC*	*ERP (EPC-1)*
Corn	1970	0.101	v	0.996	0.004	1.228	0.228
	1972	-0.996	x	1.545	-0.545	1.527	0.527
	1974	-1.994	x	1.425	-0.425	1.068	0.068
	1976	-1.213	x	1.338	-0.338	1.236	0.236
	1978	-2.492	x	1.812	-0.812	1.856	0.856
	1980	-4.453	x	2.047	-1.047	1.727	0.727
	1982	-6.611	x	3.080	-2.432	3.501	2.501
	1984	-4.143	x	1.988	-0.988	1.816	0.816
	1986	-5.445	x	3.239	-2.239	3.373	2.373
Soybeans	1970	0.356	v	0.951	0.049	1.187	0.187
	1972	2.464	v	0.854	0.146	1.173	0.173
	1974	0.260	v	0.978	0.022	0.896	-0.104
	1976	-1.070	x	1.191	-0.191	0.990	-0.010
	1978	-0.873	x	1.185	-0.185	1.404	0.404
	1980	-2.863	x	-0.413	1.143	1.515	0.515
	1982	-7.213	x	-0.956	1.956	2.315	1.315
	1984	-4.415	x	-0.584	1.508	1.584	0.584
	1986	-5.559	x	-1.015	2.015	2.845	1.845
Sugar	1970	1.127	v	0.741	0.259	1.508	0.508
	1972	2.820	v	0.582	0.418	0.972	-0.028
	1974	12.608	v	0.350	0.650	0.334	-0.666
	1976	2.675	v	0.754	0.246	0.640	-0.360

	1982	-1.020	x	1.158	-0.158	2.492	1.492
	1984	-2.878	x	1.503	-0.503	2.960	1.960
Bananas	1970	2.240	v	0.610	0.390	0.505	-0.495
	1972	1.685	v	0.666	0.334	0.535	-0.465
	1974	0.129	v	0.976	0.024	1.000	0.000
	1976	2.460	v	0.700	0.300	0.779	-0.221
	1978	1.632	v	0.832	0.168	1.034	0.034
	1980	2.127	v	0.792	0.208	0.767	-0.233
	1982	5.529	v	0.650	0.350	0.598	-0.402
	1984	2.923	v	0.785	0.215	0.548	-0.452
	1986	4.961	v	0.645	0.355	0.738	-0.262
Rice	1970	2.170	v	0.607	0.393	1.220	0.220
	1972	1.607	v	0.729	0.271	1.185	0.185
	1974	12.504	v	0.355	0.645	0.687	-0.313
	1976	1.362	v	0.868	0.132	1.688	0.688
	1978	5.501	v	0.578	0.422	1.021	0.021
	1980	4.996	v	0.659	0.341	1.382	0.382
	1982	-2.049	x	1.262	-0.262	2.394	1.394
	1984	-3.892	x	1.665	-0.665	3.103	2.103
	1986	-5.616	x	2.527	-1.527	4.564	3.564
Cut flowers (Chrysanthemums)	1982	22.966	v	0.529	0.471	1.085	0.085
	1984	15.340	v	0.577	0.423	1.463	0.463
	1986	17.368	v	0.602	0.398	0.574	-0.426
Citrus	1982	14.488	v	0.602	0.398	0.464	-0.536

(Continues)

TABLE 9.1 *(continued)*

Commodities	*Year*	*NSP*[b]	*Comparative Advantage or Disadvantage*[c]	*DRC*[b]	*Level of Comparative Advantage or Disadvantage*[d] *(1-DRC)*	*EPC*	*ERP (EPC-1)*
(Ponkan)	1984	23.198	v	0.411	0.589	0.274	-0.726
	1986	14.509	v	0.539	0.461	0.699	-0.301
Hogs	1982	14.435	v	0.703	0.297	0.323	-0.677
	1984	16.150	v	0.734	0.266	0.138	-0.862
	1986	38.828	v	0.512	0.488	0.167	-0.833

[a] NSP: Net Social Profitability.
DRC Coefficient: Domestic Resources Costs Coefficient.
EPC: Effective Protection Coefficient.
ERP: Effective Rate of Protection (ERP = EPC-1).
[b] Computations are based on shadow price of foreign exchange rate (SPFX).
[c] v: Comparative Advantage.
x: Comparative Disadvantage.
[d] The negative figures indicate a comparative disadvantage.

Source: Author's Compilation.

during the study period. Therefore, 1974 has the highest comparative advantage in rice production as compared with either the previous years or the following ones. Interestingly, the highest comparative advantage for rice production occurred during the energy and food crises in 1973–74 when the world price of rice increased from U.S.$ 227/ton in 1973 to U.S.$ 373/ton 1974. After the food crisis, the price of rice in the world market dropped to U.S.$ 255/ton in 1976 and the DRC coefficient rose to 0.868. With decreased world market prices, domestic rice production reduced its comparative advantage in 1976.

Comparative advantage in rice production, as indicated by the (1–DRC) coefficient, had a positive advantage during the period of 1970–1980 but had a negative advantage after 1980. When judged by NSP as the shadow prices of foreign exchange, the situation was similar. However, Taiwanese farmers do not necessarily conform to criteria for social efficiency. The ERP coefficients (measured by EPC minus one) have a positive advantage in the years under review, except in 1974. These results indicate that even when production efficiency is lower domestically than in foreign countries, rice can still be produced competitively by Taiwanese farmers under government protection.

Actually, rice production in Taiwan has been actively supported by the government and many public investment programs. Profits that have accrued to the farmers and society in recent years have proved the success of efforts to increase rice yield.

Citrus and Cut Flowers

As the world economy advances, both citrus and flower consumption are increasing. Since Taiwan is located in the sub-tropical region, it has the ideal climate to produce citrus and flowers. Table 9.1 shows that all DRC coefficients are below one, implying a comparative advantage for production of citrus and cut flowers.

Citrus is also subject to relatively high negative effective rates of protection due to the trade liberalization measure of lowering tariffs for citrus imports. However, when judged by the low DRC coefficients, it is clear that citrus still has potential for trade expansion. Flowers are newly developed crops in recent years. With rising levels of world living standards, flower consumption has been increasing both in domestic and foreign markets. Similar to citrus, the cut flower industry should have a great potential in the export market.

Hogs

As shown in Table 9.1, all DRC coefficients are below one for hog production. Therefore, hog production in Taiwan has a comparative advantage. However, serious environmental pollution caused by hog production has resulted in high negative effective rates of protection by the government, as shown by the negative ERP coefficients.

Economically speaking, with the value of hog production amounting to 20 percent of the total value of agricultural production, pork is the number two farm product after rice in Taiwan. It is also one of the chief staples in the Chinese diet as well as a key supplier of animal protein. Thus, hog raising is not only a vital part of Taiwan's farm economy but a major source of nutrition as well.

In summary, as shown in Table 9.1, grain production is relatively disadvantageous because its DRC coefficients are above one. Sugar production has had a comparative disadvantage in recent years. Bananas, citrus, flowers, and hogs all have a comparative advantage for production because their DRC coefficients are below unity. These four agricultural products have a fairly low domestic resource cost in earning foreign exchange. As such, they are important products for agricultural export. One the contrary, corn, soybeans, and rice clearly have a comparative disadvantage. In particular, the DRC coefficients of corn production appear to be so high that domestic production is not profitable. However, in view of the unstable world economic situation, food security is still the main consideration in Taiwan's agricultural policy. The above findings serve to illustrate why our government has offered farmers strong incentives regarding both inputs and outputs to maintain a desired level of production for rice, soybeans, and corn.

Strength of the NT Dollar and Implications for Agricultural Trade

From 1985 to 1988, the exchange rate of the NT dollar declined from NT$ 40 equivalent to one U.S. dollar to NT$ 27 to one U.S. dollar. The NT dollar appreciated more than 30 percent against the U.S. dollar during this period. The pressure to appreciate the NT dollar came mainly from the U.S.'s request to maintain the trade balance between the United States and Taiwan. The reason, as argued by the U.S., was that the low value of Taiwan currency was the fundamental cause for the U.S. trade deficit to Taiwan. Many economists in the U.S. seem to expect an expansion of Taiwan's domestic demand and imports through an appreciating NT dollar. The expansion of imports will create an environment in which the effects of exchange rate policy should reduce the trade balance.

Although it may seem both desirable and effective to use the exchange rate to correct the Taiwan-U.S. trade imbalance, it will take time for the effects of the exchange rate to work through the economic system to achieve the trade balance. Hopes for a sudden exchange rate shift to dramatically increase imports or reduce exports are not reasonable.

Actually, exchange rate movements have had some effect on agricultural trade. However, to what extent trade can be affected depends upon the price level between trading nations. As pointed out by Henneberry, *et al.*, in their study of the impact of the dollar appreciation on trade, one should differentiate between nominal and real exchange rate changes. The nominal exchange rate represents changes caused by fundamental economic forces plus the effects of inflation in trading nations. The real exchange rate represents only changes caused by fundamental economic forces after adjusting for the effects of inflation.

In this section, the Paasche formula was used to calculate the coefficient of trade-weighted foreign exchange rates for measuring the impact of strengthening the NT dollar on Taiwan's agricultural trade. Specifically, the Paasche formula is expressed as:

$$P_r = \frac{\sum_{i=1}^{n} W_{i1}\,(ER_{i1})\,(CPIV_1 / CPI_{i1})}{\sum_{i=1}^{n} W_{i1}\,(ER_{i0})\,(CPIV_0 / CPI_{i0})}$$

where P_r = index of trade-weighted foreign exchange rate,

ER_{i1} = exchange rate with trading country i in year 1,

ER_{i0} = exchange rate with trading country i in base year,

$CPIV_1$ = Taiwan's consumer price index in year 1,

$CPIV_0$ = Taiwan's consumer price index in base year,

CPI_{i1} = consumer price index in trading country i in year 1,

CPI_{i0} = consumer price index in trading country i in base year,

W_{i1} = shares (weights) of traded volume for country i in year 1.

When the P_r coefficient is equal to one, it indicates that change in the NT dollar would not have any impact on Taiwan's agricultural trade at all. When the coefficient is greater than one, it implies a comparative advantage for agricultural imports but a comparative disadvantage for exports. On the other hand, a coefficient of less than one would indicate a comparative advantage for agricultural exports but a comparative disadvantage for imports. This coefficient can effectively measure the effects of NT dollar appreciation on agricultural trade in Taiwan.

Since 1985, the strength of the NT dollar has had significant effects upon agricultural imports. Generally speaking, an increase in the value of the NT dollar would help increase agricultural imports. However, in my view, the low inflation rate against the exporting countries during the study period extenuated the impacts of the NT dollar appreciation on agricultural imports. Therefore, Taiwan's agricultural imports have not been affected too negatively (i.e., a surge in imports has not occurred) by the increased strength of the NT dollar.

Tables 9.2 and 9.3 show the nominal and real appreciation of the NT dollar for the 1984/1985 to 1987/1988 period. As is apparent from Tables 9.2 and 9.3, the NT dollar has appreciated not only in nominal terms but also in real terms against the exporting countries' currencies. However, the NT dollar's appreciation has not been as large in real terms as it has been in nominal terms. Considering corn, for example, we can see from Tables 9.2 and 9.3 that the NT dollar has appreciated 150.9 percent against the currencies of exporting countries in nominal terms in 1986/1987, while the real appreciation of the NT dollar has been only 113.4 percent in the same time period. The most striking difference between nominal and real terms is in the soybean coefficient, while the least amount of difference is found in the citrus coefficient. Generally speaking, the coefficients of real terms are less than the nominal terms, indicating that the impacts of the NT dollar appreciation upon agricultural imports are not as serious as one might expect.

Tables 9.4 and 9.5 present the nominal and real appreciation of the NT dollar for the major export crops in the period 1984/1985 to 1987/1988. As shown in these tables, rice and citrus have coefficients greater than one both in nominal and real terms in 1986/1987. That is, the NT dollar has been more expensive in terms of the currencies of the countries who import Taiwanese rice and citrus. Rice is strictly prohibited for export from Taiwan by the United States. Thus, the rice analysis is meaningless. Note that the main destinations for citrus exports are Hong Kong and Singapore. Thus, the NT dollar appreciation caused a comparative disadvantage for citrus export due to the strength of the NT dollar against the Hong Kong and Singapore dollars in 1986/1987.

TABLE 9.2 Nominal Trade-Weighted Foreign Exchange Rate Coefficient for Agricultural Imports, 1983/84 to 1987/88.

Year	*Total Imports*	*Corn*	*Soybeans*	*Sorghum*	*Pork*	*Apples*	*Citrus*
1983/84	ND	1.361064	0.912449	3.595017	1.345414	ND	ND
1984/85	ND	0.935901	3.077559	4.724248	1.517036	ND	ND
1985/86	ND	1.222460	4.342001	1.285023	1.935367	1.520977	3.245583
1986/87	2.4960539	1.509801	5.042156	1.602124	1.629966	1.704294	1.422128
1987/88	2.6690782	NA	NA	NA	NA	NA	NA

TABLE 9.3 Real Trade-Weighted Foreign Exchange Rate Coefficient for Agricultural Imports, 1983/84 to 1987/88.

Year	*Total Imports*	*Corn*	*Soybeans*	*Sorghum*	*Pork*	*Apples*	*Citrus*
1983/84	ND	1.141981	0.878295	1.540925	1.142984	ND	ND
1984/85	ND	0.870571	3.014393	1.468181	1.214884	ND	ND
1985/86	ND	1.035949	1.264571	1.182367	1.230299	1.153548	1.181916
1986/87	1.6077495	1.133748	1.258139	1.351798	1.185119	1.274611	1.278151
1987/88	1.5426281	NA	NA	NA	NA	NA	NA

ND: No data.
NA: Not available.

Source: Author's Compilation.

On the contrary, pork, bananas, and cut flowers all have P_r coefficients in both nominal and real terms less than one, showing a comparative advantage for these three crops for export. Note that these three crops were mainly exported to Japan and the Japanese yen has been stronger than the NT dollar. In this situation, even though the strength of the NT dollar increased, Taiwanese agricultural products still had a comparative advantage for export.

The differences between the nominal and real rates of trade-weighted foreign exchange indices for exports are not so significant. These insignificant differences suggest that the price structure among trading nations has less influence on trade (for export) than the changes in foreign exchange rates.

TABLE 9.4 Nominal Trade-Weighted Foreign Exchange Rate Coefficient for Agricultural Exports, 1983/84 to 1987/88.

Year	*Total Exports*	*Rice*	*Pork*	*Bananas*	*Citrus*	*Cut Flowers*
1983/84	ND	ND	1.010597	1.128206	1.230721	ND
1984/85	ND	ND	0.893671	0.892629	0.983251	ND
1985/86	ND	1.540053	0.794344	0.837855	0.963006	0.798417
1986/87	0.8788754	1.497288	0.755897	0.885660	1.671637	0.768316
1987/88	0.9728918	NA	NA	NA	NA	NA

TABLE 9.5 Real Trade-Weighted Foreign Exchange Rate Coefficient for Agricultural Exports, 1983/84 to 1987/88.

Year	*Total Exports*	*Rice*	*Pork*	*Bananas*	*Citrus*	*Cut Flowers*
1983/84	ND	ND	1.091070	1.217292	1.148616	ND
1984/85	ND	ND	0.944956	0.944627	0.962189	ND
1985/86	ND	1.213607	0.841039	0.858061	0.910611	0.843688
1986/87	0.8573334	1.319466	0.804082	0.880047	1.315952	0.816614
1987/88	0.8895787	NA	NA	NA	NA	NA

ND: No data.
NA: Not available.

Source: Author's Compilation.

Conclusions

Today, the Taiwanese economy faces a structural change from heavy reliance upon exports to greater dependence upon an expanded domestic demand. This transformation has both adverse and positive effects on different industries. Trade liberalization and the strong NT dollar have caused some agricultural products to lose their comparative advantage and to be forced out of the domestic and foreign markets.

The results of this study show that grain production has a comparative disadvantage. Corn, soybeans, and rice clearly have high domestic resource costs because their DRC coefficients are above one. However, the Taiwanese farmers do not necessarily conform to the criteria for DRC coefficients. With the government offering price incentives, the private profitability of rice, soybeans, and corn reveal only positive advantage for farmers to produce these crops at some desired level. Sugar production has had a comparative disadvantage in recent years. Bananas, citrus, flowers, and hogs have showed a comparative advantage for production because their DRC coefficients are below one.

The strength of the NT dollar has a significant influence on agricultural trade in Taiwan. However, due to the low inflation rates against other trading nations in recent years, the impacts of the NT dollar appreciation upon agricultural trade have not been as serious as the government and farmers had feared.

Taiwan has been increasing imports of food in recent years. However, agricultural policy is still based upon the principle that agriculture in Taiwan is indispensable in achieving food security and maintaining a stable ecological system. Taiwan's leaders believe that trade liberalization should be implemented in harmony with the structural improvement of domestic agriculture. Although Taiwan is not a member of GATT, its agricultural structure will continue to be improved through a gradual deregulation of trade restrictions.

With regard to future multilateral trade negotiations, Taiwan should not only concentrate on implementing agricultural policy reform in relation to trade liberalization or deregulation but should also give equal attention to the social and political impacts of agricultural trade liberalization. Non-economic factors, such as maintaining social and political stability, are very important and should not be overlooked.

Appendix

Methodology

The methodology applied in our analysis of comparative advantages follows that developed by Pearson, *et al.* For the ERP (effective rate of protection) calculation, we also consulted the book by Corden. Other references include Akrasaanee and Wattananuki, Boeha and Taylor, Balassa and Schydlowsky, Bruno, and Ethier. The methodology is explained as follows:

1. For NSP (net social profitabilities) calculation:[1,2]

$$NSP_j = (u_j - \overline{m}_j - r_j) V_1 - \sum_{s=2}^{m} \overline{f_{sj}} V_s + E_j$$

where

u_j = total value at world prices of the output of jth activity,
$\overline{m}_j$ = total (direct plus indirect) value (in foreign currency) of tradable materials used by the jth activity,

r_j = total (direct plus indirect) value (in foreign currency) of repatriated earnings of foreign-owned factors of production employed by the jth activity (including repatriated portions of the direct foreign factor costs and of the indirect foreign factor costs),

V_1 = shadow price of foreign exchange, expressed as a ratio of local currency to foreign currency,

$\bar{f}_{sj}$ = total (direct plus indirect) quantity of the sth primary domestic factor employed by the jth industry,

V_s = opportunity cost of sth factor,

E_j = externality of jth activity which may be negative or positive.

2. For DRC (domestic resource cost) calculations:

$$DRC_j = \frac{\sum_{s=2}^{m} \bar{f}_{sj} V_s - E_j}{u_j - \bar{m}_j - r_j} = \frac{DC_j}{NVA_j}$$

where

DC_j = opportunity cost of domestic resources employed by the jth activity (in domestic currency),

NAV_j = net foreign exchange earned or saved (in foreign currency), or equivalently, valued added at world prices.

3. For ERP (effective rate of protection) calculation:

(i) With one kind of tradable input, ERP can be expressed as:

$$ERP_j = \frac{PV' - PV}{PV}$$

where

PV' = value added at domestic prices

PV = value added at world prices,

Specifically, we may compute PV and PV' by the following formulas:

$$PV = P_j^f \; (1 - a_{ij})$$

$$PV' = P_j^f \; [\,(1 - t_j) - a_{ij} \; (1 + t_i)\,]$$

where

P_j^f = world price of output,

a_{ij} = ratio of cost of ith factor of production to total production cost per unit of output,

t_i = nominal protection rate of the ith factor of production,

t_j = nominal protection rate of output of the jth activity,

Therefore, ERP may be alternatively expressed as:

$$ERP_j = \frac{t_j - a_{ij} t_i}{1 - a_{ij}}$$

(ii) With more than one kind of tradable input, ERP can be computed by:

$$ERP_j = \frac{t_j - \sum_{i=1}^{n} a_{ij} t_i}{1 - \sum_{i=1}^{n} a_{ij}}$$

Notes

1. Bruno treats all locally produced inputs as nontradable.

2. Pearson, *et al.*, treats locally produced inputs as follows: (a) as tradable if they are fully traded, i.e., if the country also imports some of the inputs, or (b) as nontradable if they are not fully traded, i.e., if the country does not import any of the inputs.

References

Akrasanee, N. and Wattananukit, A., "Comparative Advantage in Rice Production in Thailand: A Domestic Resource Cost Study," *Food Research Institute Studies*, Vol. XV, No. 2, Food Research Institute, Stanford Univ., 1976.

Bacha, E. and Taylor, L., "Foreign Exchange Shadow Price: A Critical Review of Current Theories, " *Quarterly Journal of Economics*, 85, May 1971, pp. 197–224.

Balassa, B. and Schydlowsky, D. M., " Domestic Resource Costs and Effective Protection Once Again," *Food Research Institute Studies*, Vol. XV, No. 2, Food Research Institute, Stanford Univ., 1976.

Bruno, M., "Domestic Resource Costs and Effective Protection: Clarification and Synthesis," *Journal of Political Economy*, Vol. 80, No. 1, Jan./Feb. 1972.

Corden, W.M., *The Theory of Protection*, London: Oxford University Press, 1971.

Ethier, W., "Input Substitution and the Concept of the Effective Rate of Protection," *Journal of Political Economy*, Vol. 80, No. 1, Jan./Feb. 1972.

Henneberry, D., Henneberry, S., and Tweeten, L., "The Strength of the Dollar: An Analysis of Trade-Weighted Foreign Exchange Rate Indices with Implications of Agricultural Trade," *Agribusiness: International Journal*, Vol. 3, No. 3, 1987.

Pearson, S.R., Akrasanee, N., and Nelson, G., "Comparative Advantage in Rice Production: A Methodological Introduction," *Food Research Institute Studies*, Vol. XV, No. 2, Food Research Institute, Stanford Univ., 1976.

U.S. Department of Agriculture, "Government Intervention in Agriculture, Measurement, Evaluation, and Implications for Trade Negotiations," Economic Research Service, 1987.

Discussion

Zhu Daohua

I am very glad to have the opportunity to attend this symposium on China's rural development, and I want to thank the conference organizers for inviting me to serve as a discussant. I am equally delighted to have this opportunity to make a few remarks on Professor Chen Hsi-huang's chapter titled "Taiwan's Agricultural Trade and Prospect on Foreign Competition." However, because of the four decades of separation, I find myself poorly informed of the situation and realities of Taiwan's agricultural development. Therefore, I am afraid that my comments may not be faultless, and I hope that I may be forgiven.

Since World War II, Taiwan's agriculture has completed its transition from a labor-intensive to a capital-intensive operation. The increased level of commercialization of farm production, the development of the farm product processing industry, and the strengthening of economic ties with other countries, all have brought Taiwan's agriculture to a new era when it must open its domestic market to foreign goods and participate in international trade competition. In the meantime, as argued by Professor Chen, the surging NT$ (New Taiwan Dollar) has further complicated the future development of Taiwan's agriculture. Professor Chen's study is an attempt to address these problems. I find his work to be tinged with strong purposefulness and practicality. Professor Chen systematically elucidates the goals of Taiwan's agricultural policy and the relationships between domestic farm policy and farm trade policy. He applies the principle of comparative advantage to analyze the advantages and disadvantages of importing or exporting various farm products. He further investigates the impact of an appreciating NT$ upon the import and export of farm products, discusses structural changes in Taiwan's agriculture in the future, and points to possible measures to accomplish these policy goals. I have no doubt that his study will contribute enormously to the future development of Taiwan's agriculture and to the enhancement of Taiwan's competitiveness in international farm trade.

International trade should be based on an international division of labor and conducted according to the principle of comparative advantage. In the light of the requirements of net social profitability, domestic resource costs and the effective rate of protection, Professor Chen conducts a comparative advantage analysis of the foreign exchange earning ability of eight of Taiwan's farm products. Such an analysis is very useful in defining the advantage of Taiwan's agricultural production, in making rational use of resources, in readjusting production structure, and in developing the foreign trade of agricultural products. It also has certain reference meaning for studying mainland agricultural product trade. However, as is pointed out in the chapter, comparative advantage is just one of the economic principles to be followed in farm trade, but not the sole criterion. Social cost may override economic benefit in opening domestic markets and developing international markets for agricultural trade. Apart from conducting necessary microanalysis of comparative advantage for various farm products in import and export, I think special attention must be paid to the following.

First, in analyzing the costs and benefits of agricultural trade in a nation, one must take into account the position agriculture occupies and the role it plays in overall social-economic development. Full security of the food supply is the prerequisite for the successful development of the national economy and social stability. Although the proportion and impact of agriculture in the national economy differ by country and region, the state of agricultural development is directly linked with the development of the economy and society. Taiwan differs from the mainland in the level of economic development, thus one may reach different conclusions regarding trade strategies and policies. For instance, while self-sufficiency in food is a must for the mainland, this may not necessarily be true for Taiwan. Nevertheless, in view of similar population density across the Taiwan strait, the goal of obtaining food security and social stability through agricultural development should be the same. The picture of world agricultural development shows that most of the economically developed nations, whether farm product importers, like Japan and West Germany, or exporters, like the U.S. and France, all protect their agriculture to varying extents, even though they usually have technological advantage in manufacturing and industry. Thus, differences between individual nations in production structure and type of agriculture can be illustrated by comparative advantages, but the development of the entire agricultural sector can not be explicitly interpreted by this same principle.

Next, agricultural production is multisectoral, including grains, cash crops, and livestock husbandries. The production structure of agriculture is comprehensive in nature. Of course, the division of labor and product specialization are general trends in the development of modern production. It is important to note that agriculture utilizes such natural resources as land and water to carry out plant and animal production. Thus, apart from being subject to economic laws, it is subject to the laws of nature as well. Between various sectors within agriculture and between agriculture and the environment there exists an intrinsic relationship of interdependence and mutual restriction. Therefore, in adjusting the production of individual farm products, it is necessary to take into consideration the implications that such an action might have for the production structure. Only by doing so can we achieve the goals of utilizing intelligently and effectively all natural and labor resources and attaining a coordinated and balanced relationship between various sectors within agriculture, thereby securing satisfactory economic, social, and ecological benefits. So far as agricultural production is concerned, from the standpoint of the social division of labor and product specialization, comparative advantage is of vital importance. But for different farm products, comparative advantage must be examined in the context of the overall social and ecological benefits of agriculture as a whole, and even of the entire national economy.

Furthermore, the comparative advantage for agriculture is dynamic. For example, importing certain kinds of farm products may prove to be more profitable than producing them domestically. However, for the sake of protecting the interest of farmers, it is sometimes necessary to adopt some kind of protectionist policy. In my view, this is permissable. But from a long-term point of view, it is also necessary to increase as much as possible the production efficiency of farmers. In this regard, some farm products in Taiwan, such as rice and corn, are similar to those in the mainland. That is, for both sides it is essential to enhance the competitive power of these products through expanding the scale of farming operations, improving agricultural production techniques, and socializating farm production.

Finally, comparative advantage itself is a relative concept, since the comparative advantage for a country in exporting a farm product is related not only to production conditions of the exporting country, but also to those of its trade partners. Therefore, it is influenced by the world price levels, foreign exchange rates, and market conditions. There are often cases where a farm product may not have a positive comparative advantage if exported to Country A, but does if exported to Country B. In his analysis of Taiwan's trade partners, Professor Chen did not include Mainland China. In fact, there are vast vistas for farm trade between Taiwan and the mainland. A more direct trade of farm products across the Taiwan straits would likely revise the comparative advantages of certain export farm products estimated by Professor Chen.

It seems to me that in developing agricultural trade and in pushing ahead with the import and export of farm products, it is important to pay attention not only to the microanalysis of comparative advantage for farm products, but also to macro-linkages with overall economic development of the country or region, and with the development of the world economy. Only through this approach can a well-balanced and precise conclusion regarding Taiwan's trade strategies and policies be obtained.

PART FIVE

Grain and Rural Development

Temporary grain storage bins at a flour mill in China. *(Courtesy Frederick Crook)*

10

The Growing Grain Deficit and Its Implications for China's Developmental Strategy

Dwight H. Perkins

Introduction

In the late 1980s, the issue of grain once again became a topic for debate within China's policy circles. Prior to that time, the sharp acceleration in grain production between 1978 and 1984 had led some to conclude that China had solved its grain problem, at least for the time being. Grain output peaked, however, in 1984 at 407 million tons and stayed below that level throughout the next four years. From being a net exporter of grain in 1985 and 1986, China in 1987 and 1988 returned to its more normal posture of being a net importer of cereals.

The grain question, however, cannot be studied in isolation. It is part of a much broader and more important range of issues and can only be understood in that context. Grain, for example, is only one component of farmer income and in fact is now a declining factor. Rational farmers responding to market signals in the 1980s spent more time and resources on cash crops and subsidiary activities than on grain. From their point of view, there was no grain problem, only an opportunity to raise their standard of living by cultivating cotton and vegetables or by working in a local construction enterprise.

But the grain issue is also vested in the underlying question of what is the appropriate overall development strategy for China. In that context, it is useful to begin by pointing out that there are few dietary substitutes for grain. Meat, to be sure, is a major source of calories in richer countries, but meat in most such countries is a kind of processed grain. Cattle and hogs are fattened on feed grains in most of the world. Fish are less frequently fed with grain, but China does not have a large enough fishing industry for aquatic products to become a major substitute for grain. Thus, China must somehow find a way to supply its consumers with grain. The source can be domestic production or imports of cereals produced abroad, but one option that is not open is to ignore the grain supply altogether. In the analysis that follows, the first step will be to try to understand the forces at work in determining the demand for and supply of grain in China over the decade of the 1980s and in the decades that preceded the 1980s. With past performance as background, the next step will be to speculate about what is likely to happen to grain demand and supply over the next decade or two. The final step is to relate those projections into the future to the broader question of what choices China has in its overall development strategy in the coming years. To telegraph the primary conclusion of this essay, China basically does not have many viable choices when it comes to a long- term development strategy. If China wants to grow rapidly, imports will have to also grow rapidly, and one of the important components of that rising import bill will be grain. To pay for that grain and other essential imports, Chinese exports will have to increase rapidly, too.

Grain Demand and Supply in the 1980s

There is a simple equation that captures some of the critical elements determining the demand for grain.

$$\dot{D} = \dot{p} + \eta_y \dot{y}$$

where,

$\dot{D}$ = the rate of increase in demand for grain,
$\dot{p}$ = the rate of increase in population,
η_y = the income elasticity of demand for grain,
$\dot{y}$ = the rate of increase in income per capita.

The demand for grain as used here refers both to grain consumed directly and grain fed to animals and poultry and consumed indirectly.

Those who talk about the grain problem in China often concentrate on $\dot{p}$, the rate of population growth. But China's population growth in the 1980s at one point fell to 1.2 percent a year before rising to 1.5 percent toward the end of the decade. Even if China's population were to return to the 2 percent rate of much of the 1960s and 1970s (a not very likely prospect), the increase in demand for grain each year would be less than 1 percent a year above what would have occurred if the 1.2 percent rate had been maintained.

The bigger problem of rapid increases in demand for grain during the 1980s was caused, not by accelerating population growth, but by increases in per capita demand brought about by rising personal incomes. Between the end of 1978 and 1988, Chinese GNP rose in real terms at an average rate of 9.6 percent/yr. Because population grew at an average rate of 1.3 percent/ yr over the same period, per capita GNP rose at 8.3 percent/yr.

Per capita GNP is not the same thing as personal income, and we do not have direct estimates of personal income for China. We do know, however, that the share of consumption in net material product was slightly higher in 1987 (65.3 percent) than in 1978 (63.5 percent) and that personal savings rose sharply over this period. Taken together these trends suggest that personal income was growing at least as quickly as per capita GNP or at over 8 percent a year. If the GNP figure itself is distorted by unrealistic prices or unreliable statistical reporting systems in particular sectors, then GNP per capita may be a bit below the officially estimated figure, in which case the growth of personal income per person per year would also be lower. It may be better to talk about a possible range for personal income growth of 7-8.5 percent/yr than a single precise figure.

What remains is to come up with some kind of estimate for the income elasticity of demand for grain in China. In the analysis by Peng Tso-Kwei (Chapter 6), the income elasticity of demand for grain in Taiwan in the 1962-66 period was 0.5.[1] Taiwan's per capita GNP in 1986 U.S. dollars at that time was about $650. According to the World Bank Atlas, China in 1987 had a per capita GNP of only $300, but this figure reflects the peculiar features of China's official exchange rate and cannot be compared usefully with the Taiwan figure.[2] In my opinion, a figure for China roughly comparable to that for Taiwan would be in the $500 range. Perhaps then an income elasticity of demand for China would be similar to that of Taiwan in this earlier period. International experience outside of East Asia, however, would suggest a somewhat higher figure, conceivably as high as 0.7 or 0.8.

The reader can make his or her own assumptions about which of these estimates is the correct one. Here we shall present a range.

$$\text{High estimate } \dot{D} = .013 + .7 \times .083 = .071$$

$$\text{Low estimate } \dot{D} = .013 + .5 \times .07 = .048.$$

During the reform period 1978-1988, if these figures are correct, the demand for grain in China was growing at a rate of 5-7 percent/yr.

The supply side of the story is straightforward. Between 1978 and 1984, grain output in China (including potatoes and soybeans) grew at an average annual rate of 5 percent, but grain output peaked in 1984 at 407 million tons (see Table 10.1). Over the entire 1979-1988 decade, the rate of increase in grain output averaged 2.6 percent/yr.

TABLE 10.1 Grain Output in China

Year	*Amount (million tons)*[a]
1952	163.92
1957	195.05
1962	160.00
1965	194.53
1970	239.96
1975	284.52
1978	304.77
1980	320.56
1984	407.31
1988	394.00

[a]Includes soybeans and potatoes. Potatoes were converted to grain equivalents at the rate of 4 kg of potatoes to 1 kg of grain before 1963 and 5 kg to 1 kg of grain thereafter.

Sources: State Statistical Bureau, *Zhongguo tongji nianjian, 1988*, p. 248; and "The Chinese Economy in 1988," *Beijing Review*, Vol. 32, No. 6, February 6-12, 1989, p. 23.

Over the reform decade as a whole, therefore, there was a gap of more than 2 percent/yr between the growth rates of demand and supply. In the first six years of the decade, however, there was no gap at all, and that is when China switched over from being a net importer to a net exporter (see Table 10.2). In the final four years of the decade, in contrast, the demand-supply gap was more than 5 percent/yr, or conceivably as much as 80 million tons of grain, by early 1989.

How was this gap in the latter part of the period filled? To begin with, China switched from being a net exporter of 3 million tons of grain in 1985 to a net importer of over 8 million tons in 1987 and 1988, thus closing 11 million tons of the gap. We do not have data, however, that explains how the large remaining portion of the gap was covered. Presumably, there was a run-down in grain stores that had been built up in the early 1980s, but precise figures are not available. There was also pressure on grain prices. In 1986 and 1987, the free market price of grain sold rose by 42.4 percent, while the overall free market price index rose only 24.7 percent. The official price of grain was also raised; but despite that, the free market prices in 1987 were 93 percent higher than the state's retail prices. It may be that there was also some informal rationing of grain. In a matter of only 3 or 4 years, China went

TABLE 10.2 Grain Imports and Exports (million tons)

Year	*Exports*	*Imports*	*Net Imports*[a]
1952	1.53	0	-1.53
1957	2.09	.17	-1.93
1962	1.03	4.92	3.89
1965	2.42	6.41	3.99
1970	2.12	5.36	3.24
1975	2.81	3.76	0.93
1978	1.88	8.83	6.96
1980	1.62	13.43	11.81
1981	0.99	13.83	12.84
1982	0.81	16.15	15.34
1983	1.15	13.53	12.41
1984	3.19	10.41	7.22
1985	9.33	5.97	-3.36
1986	9.42	7.73	-1.69
1987	7.37	16.28	8.91
1988	7.18	15.33	8.15

[a]Minus represents net exports.

Sources: State Statistical Bureau, *Statistical Yearbook of China, 1981*, pp. 372 and 388; State Statistical Bureau, *Statistical Yearbook of China, 1982, 1984, 1986, 1988* (various issues, various pages); and General Administration of Customs, *China's Customs Statistics*, 1989, Vol. 1, pp. 21 and 30.

from a country worried about where to store all of its surplus grain to a nation that was struggling to fill a large and rising gap between demand and supply.

Why was there such a rapid change from a surplus to a deficit in the 1980s rather than in the years prior to reform? The answer is readily supplied when one looks at the supply and demand equations for these earlier

$$\dot{D} = .02 + .7 \times .023 = .036$$

In the 1960s and 1970s, GNP averaged 4.5 percent growth/yr, and the rate of state investment rose steadily, while the share of consumption fell. Thus, the personal income growth rate was below that of GNP per capita. Incomes were lower then, so the income elasticity of demand was probably higher. The population growth rate was also higher than in the 1980s, averaging about 2 percent/yr.

Grain output in at least some of these pre-reform years almost kept up with demand. Between 1965 and 1978, for example, grain output averaged an annual increase of 3.5 percent. For the entire two decades preceding reform, however, the average growth rate was only 2 percent.

The story of what happened to the demand-supply gap in these pre-reform decades is well known. In the crisis years (1959-1961), grain output fell sharply; and rations per capita, already tight, were cut to the bone in an effort to even out the deprivation. The effort was not entirely successful, to put it mildly, and tens of millions of people died from causes largely related to malnutrition. By 1965, grain output had recovered to 1957 levels in absolute but not per capita terms. In the years that followed up to 1978, domestic supplies per capita rose in most years, but so did demand. As China's foreign exchange situation improved, the demand-supply gap was filled in part by rising imports. By 1978, China was a net importer of 7 million tons of grain (see Table 10.2), but rationing of grain sales to consumers was still in force.

The absence of a severe supply-demand gap for grain before the reform period reflected in part, therefore, the fact that per capita grain output and per capita supplies were rising slowly but steadily. Of equal or greater importance, however, was the fact that personal incomes were also growing slowly. If personal incomes had risen in these early years as fast as they did in the 1980s, China would have faced a severe grain shortage much earlier.

The shortage that China did face in the late 1980s and could have faced in the early 1970s if incomes had grown faster, however, was fundamentally different in character from the shortages of the 1959-1961 period. In the latter case, the gap between supply and demand was created by a drop in supply that drove parts of China's population below a subsistence level. In the late 1980s, the gap was created more by the accelerated increase in demand. In 1987, per capita grain production was 359 kg. In 1965, by way of contrast, per capita production was 268 kg, and in 1962 it was 238 kg. At the bottom of the 1959-1961 crisis years, per capita production was even lower. The shortages of the late 1980s were a result of prosperity. The shortages of the early 1960s were the result of extreme poverty. The two situations could not be more different, and the solutions called for should also be different. We shall return to the issue of solutions below.

Before concluding this section, however, it is important to reemphasize that the data on which demand calculations are based are flawed. The elasticity estimates in particular are little more than informed guesses. Clearly, there is a need for more carefully constructed estimates of the income elasticity of demand for the consumption of grain and meat and the use of those elasticities, to come up with an overall elasticity that can be used in the basic equation presented here.

Future Demand and Supply

The Demand Side

The components of any forecast of the future of China's grain market are the same as those used to analyze trends in the past. The demand side involves forecasts of three elements: the rate of population growth, the future rate of increase in personal income or per capita GNP, and the trend in the income elasticity of demand for grain. These forecasts are speculative, but most of the plausible guesses about these trends point to a rising gap between the demand for and supply of grain.

Forecasting the future rate of population, growth presents the fewest difficulties, because there is probably only a 1 percent difference between the extremes of what is likely to occur. At the low end, China could continue for a decade and longer to pursue a one-child-per-family policy that would be effective in urban areas but less effective among farmers. Conceivably, such a policy could drive the population growth rate down to 1 percent/yr (a crude fertility rate of perhaps 17/1000, and a mortality rate of 7/1000). There is very little prospect any time soon that China can lower fertility more than this, despite the government's desire to achieve zero population growth two or three decades into the next century. At the upper end, it is conceivable that China's population could rise to the long-term levels achieved in the 1960s and 1970s of 2 percent/yr, which would imply a fertility rate of about 27/1000. Higher rates than that would imply fundamental changes in Chinese behavior back toward a preference for very large families, and that doesn't seem likely.

China's future GNP growth rate is more difficult to forecast. All one can do is establish a plausible range. The rate of growth of the past decade of a little over 9 percent/yr is the upper end of that range. In fact, the recent slowdown in agricultural output growth, the problem of controlling inflation, and the long-term impact of the political turmoil of May-June 1989 all suggest that achieving a future GNP growth rate of 9 percent/yr will be extremely difficult. In fact, a rate of 8 percent/yr may be a better guess as to the upper end of what is possible.

In establishing the lower end of this range, it is useful to keep in mind that China is likely to maintain a high rate of gross capital formation of around 30 percent of GNP. Furthermore, there is every reason to believe that China will continue the policy of openness to foreign trade and will not revert back to the autarkic policies of the 1960s and early 1970s. Thus, productivity growth should be higher than in those earlier years. Political instability over the next decade could be a negative influence, but it is unlikely that China will experience instability of the magnitude of what occured during the Cultural Revolution. Nor are there likely to be policy errors of the magnitude of those that occurred during the Great Leap Forward in 1958-1960. Overall, therefore, unless there are long-term underlying weaknesses in the Chinese economy that are not now apparent, China's economy should do at least as well as it did during the 1967-1976 period when GNP grew at a rate of around 4.5 percent/yr.

A forecast of the income elasticity of demand for grain in China in the 1990s can be based on international experience. In the calculations below, we shall use Taiwan's elasticity rate for the early 1960s of 0.5. The current mainland Chinese elasticity, as indicated above, is probably higher; and, if Chinese per capita income grows rapidly over the next decade or two, the future income elasticity of demand could fall below 0.5.

If these forecasts of the three components of the demand equation are roughly correct, they lead to the following estimates for the future growth rate in demand for grain. These estimates are presented as a range—a low and high estimate and one example of what might occur in the middle of the range.

Low estimate: $= .01 + 0.5 \times .04 = .030$
Middle estimate: $= .015 + 0.5 \times .06 = .045$
High estimate: $= .02 + 0.5 \times .07 = .055.$

The low estimate could represent what might happen if China reverts back to an inefficient controlled economy with vigorous enforcement of the one-child-per-family policy. The high estimate might be achievable if China abandons most controls and is willing to live with a high rate of inflation.

These estimates are clearly sensitive to the figure used for the income elasticity of demand for grain. If we had used 0.7 instead of 0.5, for example, the range of growth rates in demand for grain would rise to between .038 to .069. This underlines the point made above that more reliable estimates of these elasticities are needed.

The Supply Side

Is it conceivable that supplies of grain can keep up with these rates of increases in demand? There are several kinds of evidence that have some bearing on this question. There are the historical rates of growth in Chinese grain output achieved in the past; there is the experience of China's East Asian neighbors who have a land endowment and agricultural technology somewhat similar to that of the Chinese mainland; and there is information on the likely future growth of key inputs into grain output in China.

Between 1957 and 1989, Chinese grain output averaged a rate of increase of 2.2 percent/yr, but there have been periods when China has done better than that. One such period, as already mentioned, was the first six years of reform, 1979-1984, when grain grew at an annual average rate of 5 percent. But a 5 percent rate is clearly unrealistic for the long term. That rate was the product of one shot gains from the rural reforms, plus a dose of good weather in 1984. Of greater relevance is the 3.5 percent/yr rate achieved between 1965 and 1978. By 1965, Chinese grain output had recovered from the devastation of 1959-1961; and 1978 was the last year before the introduction of rural reforms, although weather was not very good in that year. Growth in the 1965-1978 period was achieved mainly by increases in inputs of fertilizer, combined with the introduction of new plant varieties and some expansion of the irrigated acreage. Since 1978, however, the average rate of increase in grain output through 1988 was only 2.6 percent.

A brief survey of the recent experience with grain production of Japan, South Korea, and Taiwan suggests that sustained increases of over 3 percent/yr may be difficult to achieve. In Japan, despite prices for rice that are several times world levels, cereal production in the latter half of the 1980s was no higher than it was in 1976, a zero percent growth rate for a decade and more. In South Korea as well, cereal production in the late 1980s was about the same as or a little below the level of the mid-1970s. Rice production in Taiwan actually fell after 1976 and the growth rate of other cereal crops was even more sharply negative. Per capita incomes in Japan, South Korea, and Taiwan, however, are much higher than on the Chinese mainland and these societies are highly industrialized and urbanized. There are factors working to slow grain output growth operating on both the demand and supply side that might not be applicable to China. The small size of the rural labor force is one case in point.

If one analyzes the input side of the Chinese agricultural output equation, however, there are not many grounds for optimism.[3] There is no better place to start than with the question of arable land. In 1957, China's State Statistical Bureau estimated that the country had 112 million ha of arable land.[4] By 1987 that estimate had fallen to 96 million ha despite two decades up to the mid-1970s when massive amounts of labor were mobilized in part to open up new arable land.[5] These estimates of arable land may not be very reliable because some new land development has gone unreported. But there is little doubt that the overall trend is negative, thanks largely to pressures from urbanization and industrialization. The pressures to convert land to nonagricultural uses can only get stronger in the decade to come.

Is it possible for China to expand the area under irrigation so that the country could expand multiple cropping and achieve higher grain yields per ha? An increase in irrigated acreage from 33 million ha to 45 million ha between 1965 and 1978 was part of the reason for the relatively high grain output growth rate of that period. But there has been no increase in irrigated acreage since 1978, and the reasons are straightforward. Most of the rise between 1965 and 1978 was achieved by digging tubewells on the North China Plain, and the potential for that kind of expansion was exhausted before 1978.

Efforts to bring irrigated water to the remaining 50 million ha of Chinese farmland are likely to prove to be extremely expensive or infeasible altogether. Rainfall over most of this acreage averages only 500 to 700 mm/yr. The Yellow River, the largest river in the nonirrigated parts of China, is one of the most silt-laden rivers in the world, comparable to the Colorado in the U.S. Removing the silt from this river by grassland development and by building dams on its tributaries and upper reaches will be extremely expensive. And the rate of flow or discharge of the Yellow River is not very large to begin with, only 6 percent of the rate of discharge of the Yangtze River. There has been talk of schemes to move Yangtze River water to the north to make up for this lack, but those schemes would also be enormously expensive.

In summary, most of the arable land of China that is going to be irrigated is already being irrigated. In fact, half of the land currently irrigated was probably under some form of irrigation by the middle of the Qing dynasty.[6] An important part of the rise in grain output both prior to 1978 and in the 1980s was the increased availability of chemical fertilizer. In 1987 China was using 20 million tons of chemical fertilizer (measured in terms of nutrient) or a little over 200 kg/arable ha (about 140 kg/sown ha). By way of comparison, Taiwan rice used around 250 kg of nutrient/ha and the highly subsidized Japanese rice farmers use over 400 kg/ha. But Chinese farmers still use substantial amounts of organic fertilizer as well. Clearly there is not a lot of room for major further increases in grain output simply by applying more chemical nutrient to the soil.

In terms of technological solutions, therefore, future growth in grain output depends on the research and development of higher yielding varieties of grain and their rapid distribution to Chinese farmers. I am completely unqualified to judge what this research potential may be. Presumably, recent political events won't make research breakthroughs of the required magnitude any easier if researchers trained abroad don't return to China to work on these problems. Even if the atmosphere in China does change enough so that these researchers do return and those in China are free to spend their time on research, the task ahead is formidable.

To what degree could improved incentives make it possible to use inputs more efficiently to achieve higher yields and output? There is some potential for institutional change that would enhance incentives. D. Gale Johnson (Chapter 3) writes about the instability of land tenure as the result of the lack of well-defined property rights.[7] Clearer property rights would give farmers a greater incentive to maintain existing irrigation systems, something many have been apparently unwilling to do in recent years. But the big increases in incentives through institutional reform have already been achieved. The return to household-based farming restored the clear connection between effort and reward that had been attenuated by the work point system of the communes and production teams. For the future, there is more of a danger that this connection between work and reward will be eroded than there is a prospect that it will be improved.

Could higher grain price make a substantial contribution to the attainment of higher grain output? To do so, these higher prices would presumably have to lead to a shift in resources away from other crops toward grain. The data on sown acreage in grain gives one some idea of what is possible. In 1987, 77 percent of the sown acreage in China was in grain. The highest percentage of acreage in grain, 87 percent of the total, was achieved in 1952 and 1962. These latter years were ones in which Chinese farmers were struggling for survival and did everything in their power to ensure that their families received at least minimum requirements for nutrition. The 87 percent figure can thus be treated as an upper limit on how much land could be shifted back into grain by higher grain prices. Given that Chinese farmers are not now anywhere near the subsistence level for the most part, a realistic shift of land back into grain would be much smaller.

It would take an extraordinarily high grain price to achieve an increase in grain output of 10 percent by that means. The current official average grain purchase price of $135/ton is roughly at world prices, so there is little potential for increase without following the Japanese example of keeping prices well above the world level.[8] China is already selling grain at or below cost, depending on which market it is sold on, so that further increases in the purchase price would either have to come out of the government budget or the budget of individual consumers. The government was already strained in the early 1980s by the cost of agricultural subsidies and is not likely to want to go a long way toward even larger subsidies.[9] Higher grain prices to consumers risk further alienation of an urban population that is not happy with the government as it is. In short, there is not a lot of potential for expanding grain output by paying farmers higher prices for their output.

Could the government accomplish by force through compulsory quotas what it might be unwilling to do with higher purchase prices? In effect, this would be a reversion to policies of an earlier era, when slogans such as "take grain as the key link" were popular with the leadership. It is an open question whether China's government today has either the will or the power to reestablish grain quotas for China's farm households. But even if the government does have the will and the power to do so, what would be gained? The government might be able to force a shift of acreage to grain from 77 percent up to 82 or 83 percent of the total. Forcing a shift in other inputs such as chemical fertilizer would be more difficult to achieve, because it would be almost impossible to monitor. Thus, the use of compulsory grain quotas might raise grain output by 5 percent. But that assumes that farmer incentives to work efficiently on grain would not be much affected, and that is highly unlikely. If farmers shift land into grain but turn all of their efforts to other crops not so tightly controlled (a prospect that is virtually a certainty), then grain output might actually fall despite the increase in sown acreage.

In summary, there are no easy methods for raising the rate of growth in Chinese grain output. Research into new higher yielding plant varieties is central to whatever success is to be achieved. Higher prices for grain and more secure property rights will help. Further increases in the use of chemical fertilizer, better machinery for pumping water, and increases in other inputs, will play a positive role, even if subject to diminishing returns. What will all of these measures imply for the growth rate of grain output? It doesn't seem likely that China can match the 3.5 percent rate of the 1965-1978 period. A decline to a growth rate of 1 percent/yr would have grave implications for a rapidly developing economy, so one assumes a major effort will be made to do better than that. Thus, one is left with a likely agricultural growth rate of 2 or 3 percent/yr. If grain output growth is not in this range, it is more likely to be below 2 percent than much above 3 percent for any period as long as a decade.

The Widening Gap

If these projections of the demand for and supply of grain are roughly accurate, then the implications are quite clear. If China can keep grain production increasing at 3 percent/yr, then China can keep this demand supplied from domestic sources if the overall growth rate in personal incomes is held to 4 percent /capita /yr and the population growth rate is reduced to 1 percent/ yr. Any rate of growth of personal incomes above this level, with or without a higher population growth rate, will lead to a widening gap between grain demand and supply.

How large could this gap become? If we take our mid-range estimate of the growth in demand of 4.5 percent/yr and a supply growth rate of 3 percent, the difference is 1.5 percent/yr or a gap that widens by 6 million tons of grain each year. The next question then is how will this gap be filled?

If the Chinese government is determined to close this grain deficit by relying only on domestic sources of grain, there is only one way over the long run that this can be done. The GNP and personal income growth rate will have to be held to levels little or no higher than what was achieved in the pre-reform period, and large investments will have to be made in agriculture and agricultural research. Over the short run of several years, the government can also close the gap by reimposing an increasingly rigorous system of food rationing, but rationing will have several disincentive effects if it alone is used to close the gap.

The other solution, of course, is to use imports of grain to make up the deficit without a resort to rationing. These imports will have to rise steadily year by year and will make the 1988 net imports of 8 million tons seem small indeed. How large these imports will have to be depends on too many assumptions to be precise. If incomes per capita grow at 5 or 6 percent/yr and production only at 2 percent, then imports could be as high as 80 to 100 million tons by the year 2000. If demand growth is lower, if production increases faster, and if there is some increase in rationing and restrictions or price incentives to discourage converting grain into beef and pork, then it may be possible to hold net imports by the year 2000 to 30-40 million tons and still enjoy a fairly high growth rate.

Could China afford to import 40 million tons of grain each year or twice that amount? At $130/ton, the cost in foreign exchange would be $5 billion to $10 billion in foreign exchange each year.[10] But then prices are not likely to stay at current levels if China increases its purchases by figures of this magnitude, so the foreign exchange requirements for even 40 million tons may approach $7-$10 billion, depending on what happens to demand and supply in the rest of the world.

These are large figures, but so are the figures for China's total exports and foreign exchange earnings. In 1975, China's total exports were only $7.26 billion, and they were at $9.75 billion in 1978. After 10 years of reform, in contrast, Chinese exports in 1988 reached $47.6 billion. In addition, China was receiving another $8 billion in credits and direct investment. In this context, China's actual cereal imports (net) of $706 million are barely worth noticing. They certainly were not a significant cause of shortages of priority imports for other sectors.

Even at today's export levels, China could afford to import (net) 30-40 million tons of grain if prices stayed down. Other sectors would have to cut back, but China would still have nearly $50 billion in foreign exchange each year to meet the needs of these other sectors. If exports grow at 9 percent / yr, a figure that is well below the 17 percent/annum rate of the previous post-reform decade just ended, then export earnings 10 years from now will be over $100 billion, and an amount 10 times what is currently spent on grain should be feasible without sacrificing much growth in the economy as a whole. In short, as long as China maintains a successful outward-looking policy in the trade sphere, increasing imports of grain should not pose a serious problem. If export growth were to slow markedly or stop altogether, however, China would probably have to markedly slow its overall growth rate to be able to avoid more imports of grain than the country could afford.

In many respects, China over the past decade has been following a development strategy with many features in common with the strategies of its East Asian neighbors. Two of the essential features of this strategy were a high rate of growth in the export of manufactures and a rapid increase in imports of food. But China cannot afford to go as far in terms of dependence on grain imports as its East Asian neighbors. Taiwan, for example, imported $880 million worth of wheat, corn, and soybeans, or $45 per capita. If the Chinese mainland were to import the same dollar amount per capita, the total value of grain and soybean imports would come to $45 billion, a figure China cannot afford now or anytime in the foreseeable future. South Korea's total food imports, including live animals, amounted to $54 per capita.[11]

Thus, a rapid growth strategy for China implies an outward-looking trade strategy because, among other reasons, that is how China will be able to afford the rising grain imports that will be needed to sustain the strategy. But China must also maintain a significant rate of growth, not only in the value of agricultural output in general, but in grain in particular. To achieve that goal, China will have to invest heavily in agriculture in the years to come, much more heavily than has been the case in the 1980s.

Notes and References

1. Peng Tso-Kwei, "Price, Income, and Farm Policy in Taiwan," (Chapter 6, this book).

2. The World Bank, *The World Bank Atlas, 1988* (Washington: The World Bank, 1988) p. 6. China's official exchange rate is overvalued in the sense that demand for imports at that rate greatly exceeds what China is able to supply in terms of finding the necessary foreign exchange, but this excess demand is in turn a product of a socialist shortage economy where firms face "soft budget constraints" and related phenomena typical of this kind of system. If China were to succeed in establishing a true market system, the current exchange rate would probably be seen as markedly undervaluing the Chinese renminbi.

3. The issues discussed in the paragraphs that follow are discussed at greater length in Dwight H. Perkins and Shadid Yusuf, *Rural Development in China* (Baltimore: John Hopkins Press, 1984).

4. State Statistical Bureau, *Weidade shinian*, p. 128.

5. State Statistical Bureau, *Zhongguo tongji nianjian*, 1988, p. 6.

6. This point is made at greater length in Dwight H. Perkins, *Agricultural Development in China, 1368-1968* (Chicago: Aldine, 1969), Chapter IV.

7. D. Gale Johnson, "Economic vs. Noneconomic Factors in Chinese Rural Development." (Chapter 3, this book).

8. The $135 is obtained by dividing the average state purchase price for all grains in 1987 by the official *yuan*-dollar exchange rate of .373 *yuan* to 1 dollar. World prices

in grain are reported separately by crop, so it is not possible to compare these in any precise manner with this Chinese figure.

9. See Nicholas Lardy, *Agriculture in China's Modern Economic Development* (Cambridge: Cambridge University Press, 1983) pp. 192-195.

10. The unit value of Chinese imports of cereals in 1988 was $123 /ton based on China's customs statistics.

11. These figures were derived from Council for Economic Planning and Development, *Taiwan Statistical Data Book, 1987*, p. 240, and Economic Planning Board, *Major Statistics of Korean Economy, 1989*, p. 208.

About the Editors and Contributors

PETER CALKINS, a professor of agricultural economics at Université Laval in Quebec City, has always had a deep interest in Chinese economic development. He earned his B.A. and M.A. degrees in Chinese literature from Harvard University in 1968 and 1971, and his Ph.D. in agricultural economics from Cornell University in 1976. Of a total of more than 10 years of professional experience in Asia, he spent four (1968-70 and 1973-75) in Nepal conducting research and extension on agricultural production systems. Between 1976-78 he returned to Taiwan, where he had studied at Tunghai University in 1966-67, as Chief Economist of the Asian Vegetable Research and Development Center. In 1982-83 he represented the Agricultural Development Council as a visiting professor at Northwest Agricultural University in Yangling, Shaanxi, where he gave courses on farm management and computer programming. For five years (1984-89), he served as the Chair of the China committee of the American Agricultural Economics Association. He has authored or coauthored over 40 articles and 4 books on comparative economic systems, economic development, agricultural production, and research methodology.

HSI-HUANG CHEN is a professor and Chairman of the Department and Graduate Institute of Agricultural Economics, National Taiwan University. He received his B.S. degree at National Taiwan University in 1959 and his M.S. and Ph.D. degrees at the University of Georgia in 1971 and 1974, respectively. He was a senior specialist at the Joint Commission on Rural Reconstruction during 1975-79, and later became the Chief of the Economic Research Division at the Council for Agricultural Planning and Development in Taiwan. Dr. Chen participated in a number of international cooperative research projects. In 1975, he was invited by the Asian Productivity Organization to carry out a research survey, "Taiwan's Food Processing Industry." He was commissioned by the Council for Asian Manpower Studies to conduct a project on "Agricultural Incentives and Implications on Employment" in 1977. During the last 20 years, his research activities have dealt with the economic and social aspects of rural Taiwan, such as food demand, agricultural labor mobility, small farms, and rural development problems and policies.

WEN S. CHERN is a professor of agricultural economics at The Ohio State University. He devoted his earlier research to energy economics and energy demand estimation while working as an economist and group leader at Oak Ridge National Laboratory during 1974-81 and as a senior economist at Lawrence Livermore National Laboratory during 1981-83. He extended his research to consumer and consumption issues while he was on the faculty (associate professor in 1983-86 and full professor in 1986-87) at the University of Maryland, College Park. His current research covers food consumption and food demand estimation in Taiwan, Mainland China, Japan, Korea, and the United States, as well as energy demand assessment in LDCs. Currently, he serves as Chair of the China Committee of the American Agricultural Economics Association. He earned his B.S. degree in agricultural economics from National Chung Hsing University in 1964, his M.S. in agricultural economics from the University of Florida in 1969, and his M.S. in statistics and Ph.D. in agricultural economics from the University of California at Berkeley in 1971 and 1975, respectively.

ZHU DAOHUA, born in 1930, graduated from the Department of Agricultural Economics of Nanjing Agricultural University in 1953. From 1953 to the present, he has been working at Shenyang Agricultural Universiy. He is currently President of the university, a professor of agricultural economics, and a tutor of doctoral students. He also holds a concurrent post of Vice Chairman of China's Agricultural Economics Association. He once won the title of nation's outstanding contribution expert. His research areas are mainly in rural development and world agriculture. Topics of his published works include: socialist agricultural economics, foreign agricultural economics, agricultural modernization, agricultural commodity, rural property economics, and agricultural and rural development strategies.

WEN-FU HSU is a professor in the Department of Agricultural Economics, and, concurrently, Dean of General Affairs at National Taiwan University, Taipei, Taiwan. Professor Hsu received his B.S. degree in Agricultural Economics from National Taiwan University, and his M.S. and Ph.D. in Agricultural Economics from Pennsylvania State University in 1961 and 1966, respectively. Dr. Hsu's teaching and research interests are in the areas of agricultural marketing and agricultural policy. His recent research includes marketing cost and margin of fruits and vegetables, changes in the trend of agricultural wholesale marketing system under the changing structure of food retailing, future agricultural policy in the trade liberalization era, and the feasibility of establishing a marketing board in Taiwan.

D. GALE JOHNSON is the Eliakim Hastings Moore Distinguished Service Professor of Economics Emeritus at The University of Chicago. He received his B.S. degree at Iowa State College, his M.S. at the University of Wisconsin, and his Ph.D. at Iowa State College. He began his career as a research associate at his alma mater in 1941 and became assistant professor of economics a year later. In 1944 he joined the faculty of the Department of Economics at The University of Chicago and in 1954 received his full professorship. In 1960 he was named Dean of the Division of the Social Sciences and served in that position until 1970. He was Chairman of the Department of Economics from 1971-1975, a position to which he returned in 1980 and held until 1984. He was Acting Director of the University Library from 1971-1972; Director, Office of Economic Analysis from 1973-1975; Vice President and Dean of Faculties, January 1975 to December 1975; and Provost, December 1975 to June 1980. He is a Fellow of the American Agricultural Economics Association and of the American Academy of Arts and Sciences. He is the author of *The People's Republic of China, 1978-1990*, a monograph on the economic reforms in China, as well as *World Agriculture in Disarray* (revised edition). Both will be published in 1991.

JUSTIN YIFU LIN is a former Deputy Director of the Development Institute, Research Center for Rural Development (RCRD) of the State Council, and is currently with the Research Department for Rural Development, Development Research Center of the State Council. He is also concurrently an associate professor teaching in the Department of Economics, Beijing University. Dr. Lin received his Ph.D. in Economics from the University of Chicago. After graduation, he conducted research at Yale University and also at the World Bank. Dr. Lin was a visiting associate professor in the Department of Economics, University of California, Los Angeles, in the fall of 1990. His major area of research includes China's rural reforms, institutional changes, agricultural productivity, and policy analysis.

YU-KANG MAO is a counselor with the Council of Agriculture, Executive Yuan, Taiwan. He was an instructor, associate professor, and professor in the Department of Agricultural Economics, National Taiwan University from 1962 to 1972. He became the Director, Rural Economics Division, Sino-American Joint Commission on Rural Reconstruction (JCRR) in 1973 until its termination in 1980. Then he served as the Director, Department of Planning and Economics, Council on Agricultural Planning and Development, Taipei, 1980-1987. During this period, he was also the Director of the Land Reform Training Institute. In addition, Dr. Mao was a visiting professor at the University of Malaya in 1972 and at Bonn University in 1988. Dr. Mao received his B.S. in Agricultural Economics from National

Taiwan University in 1955, his M.S. and his Ph.D. in Agricultural Economics from Cornell University in 1962 and 1968, respectively. Dr. Mao's teaching and research interests are in the areas of agricultural development, agricultural policy, and land economics.

TSO-KWEI PENG is a professor and Director of the Graduate Institute of Agricultural Economics at National Chung Hsing University, Taichung, Taiwan. He received his Ph.D. in Agricultural Economics from the University of Illinois, Champaign-Urbana in 1981. He was a visiting scholar in the Department of Agricultural Economics at the University of California-Davis from 1988 to 1989. Dr. Peng's teaching and research interests are in the areas of agricultural policy, agricultural trade, and agricultural prices. His published work has covered such topics as price stabilization policies and their impact on developed countries, food consumption pattern change, Taiwan agricultural trade policy, interrelationship between agriculture and industry, Taiwan's part-time farming, and Taiwan agricultural products intra-industry trade.

DWIGHT H. PERKINS is Director of the Harvard Institute for International Development and H. H. Burbank professor of political economy at Harvard University. He is the author, coauthor, or editor of 11 books and over 60 articles on the economic history and development of East Asia, particularly China and Korea, and on general economic development issues. He has served as an economic consultant to the governments of Malaysia and China, to the Korea Development Institute, to the Ford Foundation, and to the World Bank, among others. He received his undergraduate degree from Cornell University in Far Eastern Studies in 1956 and his graduate degrees in Economics from Harvard University in 1961 and 1964.

NIU RUOFENG is a professor in the Institute of Agricultural Economics (IAE) at the Chinese Academy of Agricultural Sciences (CAAS). He received a Bachelor's Degree in the Department of Economics in Leningrad Agricultural College of USSR. He was Deputy Director of IAE from 1981 to 1982, and then Director from 1982 to 1990. He is a tutor of doctoral students and offers economic lectures for the Graduate School of CAAS. He is Deputy Chairperson of Chinese Agricultural Economics Association and chief editor of the journal titled *Problems of Agricultural Economics*. His major research interest is in the following three areas: subject development of agrotechnic economics and the evaluation of agricultural research results, agricultural and food development, and the relationship between agriculture and industry. He has published many papers in these research fields. His major works include the following titles: The Conception of

Economic Effect of Agricultural Production and the Methods of its Calculation; The Evaluation of Agrotechnic Research and Its Utilization; Chinese Agriculture in 2000; The Research on Relationship Between Agriculture and Industry. He devotes himself to the exchanges and cooperation between eastern countries, and between Mainland China and Taiwan.

FRANCIS C. TUAN is the leader of the China Section, Agriculture and Trade Analysis Division, the Economic Research Service (ERS) of the United States Department of Agriculture (USDA). He received his B.S. degree in Agricultural Extension and his M.S. degree in Rural Sociology from National Taiwan University. He then received his M.S. and Ph.D. in Agricultural Economics from the University of Florida and the University of Illinois, respectively. Dr. Tuan was a visiting lecturer at University of Illinois in 1978 and 1979. He joined the ERS, USDA, in 1979 and has since monitored and studied China's agriculture. Dr. Tuan's recent publications cover China's agricultural planning and statistical systems, livestock sector, grain and oilseed crop production, and agricultural trade and policy. His research interest is in the areas of production economics, rural development, and international trade. Currently, his major research is centered on China's agricultural trade and the country's regional crop production and distribution.

GUANGSEN WANG, Ph.D. Cornell University, professor of agricultural economics at the Northwest Agricultural University, was Luce visiting professor at the Food Research Institute, Stanford University during 1983-84 and President of the Xian Statistical College from 1985 to 1989. Dr. Wang's recent research interests center in the agricultural adjustment problems in the course of economic development, with special reference to countries with very high man-land ratios, such as India, Japan and China. He conducted a three-year project on rural development in China funded by the Ministry of Agriculture and the State Education Commission. The findings were summarized in *Structural Transformation and Rural Development*, a book published by the China Finance and Economic Press in 1990.

CARSON KUNG-HSIEN WU is President of the National Pingtung Institute of Agriculture (1984-). He received his Ph.D. in Agricultural Economics from the University of Minnesota in 1968. He acted as a Senior Economist at the Economic Cooperation Centre for the Asian and Pacific Region (ECOCEN), ASPAC Bangkok, Thailand, from 1972 to 1975. He was Director of the Research Institute of Agricultural Economics at National Chung-Hsing University from 1975-1979 and Director of the Food and Fertilizer Technology Center for the Asian and Pacific Region (FFTC), ASPAC Taipei City, R.O.C., from 1979-1982. He specializes in economic evaluation of development projects, soil and water conservation of slopeland, and resource economics.

AN XI-JI is a professor of economics in the College of Agricultural Economics and Management at Beijing Agricultural University and was Chair of the IAAE Committee in China. He offered courses on theory of microeconomics, theory of macroeconomics, agricultural production economics, and agricultural development economics for graduates in the 1980s. His major interest is in economic reform and agricultural policy in China. He presented a paper titled "Pricing System Reform for Agricultural Products and Price Policy Adjustment in China, 1979-1984" (invited) at the IAAE IXX Conference in Malaga, Spain in 1985; presented a paper on "The Alternatives and World Impacts of Food Policy in China" at the 1988 World Food Conference in Des Moines, Iowa; and gave a short review of Ralph R. Johnson's policy speech at the conference. He is the author of *Agricultural Production Economics, Theory and Application in China,* and several other books.

SHIN-HSING YEH is a research fellow at the Chung-Hua Institution for Economic Research in Taipei, Taiwan. He received a B.A. in cooperative economics from National Chung-Hsing University (NCHU) in Taipei in 1972, an M.S. degree in Agricultural Economics from NCHU in Taichung in 1974, a Ph.D. degree in Agricultural Economics from National Chung-Hsing University in Taichung in 1980, and did postdoctoral study at the University of Wisconsin at Madison from 1982 to 1983. Dr. Yeh's teaching and research interests are in the areas of Mainland China's agriculture sector, natural resource development, environmental economics, and comparative cooperatives. Most of Dr. Yeh's research focuses on Mainland China's agriculture sector.

LIU ZHICHENG received his B.S. in Agricultural Economics from Chunbei University in Sichuan province in 1951. He then received his Ph.D. in Agricultural Economics and Management from the Graduate School at Soviet Leningrad Agriculture College in 1958. From 1958 to1982, Professor Liu was Chief of the Agricultural Economics Research Department, and later was the Director of Agricultural Economics Research Institute of Chinese Academy of Agricultural Sciences (CAAS). From 1982 to 1990, he was a Vice President and, later, the First Vice President of CAAS. Professor Liu's research area has been related to agricultural economic policy and macroeconomics. He coordinated several national key projects and won the first- and second-class awards of the Nation's Science and Technology Progress. Professor Liu, as editor-in-chief, published three famous works, titled *Research on China's Agriculture, Research on China's Food,* and *Research on*

Economies of Scale in Agriculture. Presently, Professor Liu is a Vice Chairman of the Academic Committee of CAAS and the First Vice Chairman of the Chinese Society of Agricultural Economics. He is also a member of the National Committee of Agricultural Regionalization, Science and Technology Committee, Ministry of Agriculture, and is a member of the editorial board of the journal, *Problems of Agricultural Economics.*

Index

Activities,
sideline, 14, 27, 35, 122, 139
Advisors,
foreign, 7
Agents,
state commercial, 126-127, 135
Agriculture, 61 (note 3), 67, 84, 99, 139, 149, 166, 182, 200;
Council of, 74, 99;
development of, 67-68, 73, 85, 97, 108, 139, 163, 183;
gross output of, 26;
growth of, 12, 16-17, 26, 43, 106, 108, 121, 198;
organization of, 1, 14, 33, 36, 45;
output of, 16-18, 28, 108, 121, 195;
policies of, 20, 100, 154, 164;
production of, 12, 18, 76, 86, 87, 99, 103, 107, 113, 121, 137, 174, 183;
production costs of, 99;
productivity of, 20, 47, 123, 154;
research in, 51, 81-82, 85, 89-90;
sector, 7, 84, 93, 137, 164;
structure of, 137, 147, 160, 179
An, Xi-Ji, 94-96, 136
Animals,
draft, 18
Assistance,
foreign, 7, 74

Bananas, 83, 165, 169
Barker, Randolph, 51
Binswanger, Hans, 47
Brigades, 14, 51;
leaders of, 32, 36
Buck, John L., 19
Buddhism, 10
Budget, government, 197

Calkins, Peter, 94
Capital,
current, 106, 109, 121;
construction, 18
Capitalism, 5, 7
Carter, C.A., 20
Cereals, 150, 159 (note 2), 166, 187, 201 (note 10)
Chang, H.T., 111
Chang, Pei-Chen, 138
Chen, Hsi-huang, 147-148, 182
Chen, Y.E., 105-107, 113
Chern, W. S., 145
China, Hong Kong, 145, 176;
Statistical Yearbook, 62 (note 12)
Chinn, Dennis L., 116
Christensen, L.R., 107
Citrus, 79, 82, 173-174, 176
Climate, 6, 173
Coefficient, Gini, 96, 116-117
Collectives, 126;
ownership of, 124
Communes, 13-15, 27, 29, 44, 51, 140
Communism, 5;
Chinese Communist Party, 33, 37
Comparative advantage, 19, 52, 111, 147, 161, 168-169, 173-179, 184
Competitiveness, 115, 156, 161, 164
Confucianism, 10
Consumption, 30-31, 103, 150, 173, 189, 191;
domestic, 138, 147, 150, 168;
final, 111
Contract, government, 169
Cooperatives, marketing, 8, 79, 85, 127;
supply, 8, 127
Corn, 20, 51, 82, 158, 165, 168, 174
Corruption, 8

Cost,
function of, 107-108;
domestic resource, 174, 180
Cotton, 14, 17, 20, 23 (note 46), 51, 127, 130, 134, 146, 150, 155
County,
agricultural technical extension organizations, 90;
farmers' associations, 75;
leaders of, 25;
Shunyi, 36;
Zouping, 35
Credit, 12, 13, 18, 71, 75, 199
Crops, cash, 139, 168, 187
Culture, 84;
heritage of, 5;
tissue, 82

Dairy products, 83, 103, 127
Demand,
elasticity of, 161, 167;
function, 58;
income elasticity of, 103, 161, 188-189, 191, 193-194;
input, 62 (note 13)
Development,
economic, 2, 5-6, 67, 71, 86, 99-100, 116, 138, 147, 173;
equitable, 15;
industrial, 68, 99;
rural, 1, 6, 12, 15, 52, 97, 113;
stages of, 8, 163
Discrimination, 32
Disparities, rural-urban, 120
Distribution, 8, 72, 77, 96, 113, 116-117, 132
Dumont, René, 15

Economy,
control, 126-127, 136;
diseconomies, 97;
growth of, 33, 94, 103, 106, 116;
market, 44, 47, 128, 132, 134;
mixed, 94, 128, 132, 134-135;
of size, 97, 140;
planned, 126, 136
Editorial Board of the Almanac of China's Foreign Economic Relations and Trade, 157
Editorial Board of the Series of China in the Year 2000, 158
Education, 71, 97
Efficiency,
allocative, 93, 97, 138;
economic, 112;
technical, 54
Egalitarianism, 10, 15, 137
Eggs, 14, 77, 103, 127, 149
Employment, 29, 36, 96, 106, 116
Engel's Law, 103
Enterprise,
private, 11;
rural, 37
Equalitarianism, 137
Equity, 97
Exchange,
foreign, 84, 147, 159, 161, 175, 180, 199;
rate of, 164, 175, 189
Exports, 20, 61 (note 3), 145, 150, 153, 156, 158, 165-166, 168-169, 176, 199;
value of, 149-150, 156, 158
Extensification, 17
Extension, 10, 13, 18, 70, 73, 76, 83-84, 90

Factories, 77, 113
Factors,
noneconomic, 25, 179
Falcon, W.P., 103
Family,
nonfarm, 115
Farm,
expenditures, 107;
family, 113, 122;
income, 96, 99, 112, 115-116, 118, 122, 137;
labor, 106, 109, 113;
land, 35, 106, 113, 196;
population, 122;
size, 67, 76, 95, 109, 111-112, 116-117;
wages, 111

Farmers, 8, 18, 36, 56, 95, 100, 113, 163;
associations, 13, 75-76;
part-time, 36, 118;
Taiwanese, 147, 173, 178
Fascism, 15
Feder, Gerson, 18
Flowers, cut, 173, 177
Food,
balance sheet, 103;
demand, 61 (note 3), 94, 103;
processing, 113, 153, 163;
security, 148, 164, 174
Freedom, 31, 74
Fruits, 79-80, 82, 156

Geography, 6
GNP, 17, 20, 67, 157, 189, 191, 193
Government,
centralization of, 33;
decentralization of, 33, 71, 127
Grain,
coarse, 158;
demand for, 154, 188, 189;
import of, 155;
output of, 20, 26, 130, 153, 187, 189, 192, 194-195, 198;
price of, 26-27, 197;
production of, 36, 153-154, 174, 178, 192, 195;
rationing of, 190, 192;
storage of, 76;
supply of, 188, 193, 198;
trade of, 127, 153
Great Leap Forward, 13-14, 153
Growth rates, 67, 103, 190

Hayami, Yujiro, 47, 49, 55, 95
Herdt, Robert W., 51
Hicks, John R., 47, 49, 55, 108
Ho, S.P.S., 116
Hogs, 130, 168, 174, 150
Holding size, 22 (note 13), 97
Household Responsibility System 27-29, 35-36, 126
Housing, 30-31
Hsiao, Cheng, 58
Hsu, Wen-fu, 64
Huai River, 6
Hypotheses, 15, 44, 56-57
Ideology, 7, 9, 52;
factors of, 30
Imports, 20, 145, 148, 150, 153, 166, 199-200
Income,
disposable, 154;
distribution of, 97, 116-117;
net cash, 129;
nonfarm, 96, 112, 115, 117, 122, 137;
personal, 189, 191, 193, 198;
policies of, 103;
real, 30-31;
quintile, 15;
sources of, 115, 122, 137;
variations of, 123
Independence, 11, 31, 85
Index,
diversification, 95, 113;
multiple cropping, 19, 95, 113
Industry, 12, 50, 60 (note 1), 67, 100, 156;
rural, 95, 113, 137
Inflation, 17-18, 176
Infrastructure, 16, 19, 113
Innovation,
mechanical, 95
Institutions, 29, 52, 56, 64, 81, 89-90
Inputs, modern, 8, 16, 51;
structural, 121;
subsidy, 137;
traditional, 8, 109
Investment,
capital, 20, 79, 116;
government, 18, 31, 139, 155;
private, 18;
production, 37, 50;
state, 29, 155, 191
Irrigation, 6, 13, 67, 77-79, 196;
Commission for, 13
Japan, 7, 16, 20, 33, 111, 145, 157, 165, 177, 195;
Jiang, Qing, 10
Johnson, D. Gale, 1, 30, 33, 103, 196

Johnson, Glenn L., 137
Joint Commission on Rural
Reconstruction (JCRR), 68, 70-74, 85

Kao, Chang, 16, 137-138
Kuomintang, 11, 13

Labor, 12, 14, 16, 19, 36, 47, 56, 86, 90, 106, 137, 157;
intensity of, 122;
nonfarm, 115;
productivity of, 17, 95, 111, 118;
rural, 14, 50
Land, 6, 30, 33, 35, 47, 51, 52, 56, 67, 113, 126, 132, 197;
arable, 6, 10, 132, 195-196;
coastal, 31, 65, 84, 156;
intensity of use, 122;
land-to-the-tiller program, 12;
man-land ratio, 123;
private ownership, 124
Lau, Lawrence, 107
Lay, Tzyy-Jane, 139
Li, Choh-ming, 113
Lin, Biao, 10
Lin, J.R., 107-108
Lin, Justin Yifu, 2, 28, 43, 47, 51-52, 60
Liu, Shao-qi, 10
Liu, Zhicheng, 44
Livestock, 18, 68, 83, 139, 168
Lu, Empress, 10

Ma, Nathaniel, 18
Management, 34, 51, 78, 126, 156
Mao, Yu-kang, 44
Mao, Zedong, 6, 9-10, 14, 16, 19, 31
Marketing, 12-13, 17, 76, 97, 100, 127, 165;
channels of, 80;
system of, 27
Markets, factor, 49, 54;
foreign, 107, 148;
free, 10, 100, 127, 190;
international, 27, 161;
world, 146, 150, 156, 158-159, 164, 169
Marxism, 6-7
McMillan, John 28, 52
Meats, 148
Medical services, 29
Methodology, 64, 179
Millet, 84
Mobilization,
mass, 50
Modernization 100, 116
Monopoly, 126-127
Mu, Yuan, 32
Mushrooms, 82, 166

Niu, Ruofeng, 18, 20, 147-148

Oil,
edible, 155, 158, 166
Ong, S.E., 112
Oshima, Harry T., 116
Organization,
adjustments to, 99, 111;
agricultural, 33, 36;
changes in, 18, 97, 178;
mass, 8;
restructuring of, 126-127;
rural labor, 14

Payments,
direct, 105
Pearson, S.R., 103, 167, 179
Pendulum, 10, 14, 94
Peng, Tso-kwei, 93-96, 106-108, 111, 113, 120-121
Perkins, Dwight, 1, 16-17, 59
Philosophy, 16
Plan, 12;
evasion of, 97, 134-135
Policy,
cost of, 106;
development, 6;
fiscal, 17;
macroeconomic, 147;
monetary, 17
Population, 6, 11, 50, 60 (note 3), 67, 138, 154;
growth of, 26, 103, 188, 191, 193, 198;
nonfarm, 118;

rural, 30, 32
Pork, 146, 150, 174, 177
Poultry, 14, 127
Price, 16, 18, 26-28, 37, 95, 100, 106, 130, 153, 164-165, 173, 190;
controls, 10, 26;
distortions, 94, 120;
factor, 47, 107, 118;
fluctuations, 167;
policies, 105, 120, 137, 166;
procurement, 134, 137;
real, 27;
retail, 26, 28, 190;
social, 97;
stabilization, 105;
supports, 105, 120-121, 169
Production, 14;
costs, 99, 108, 164, 169;
function, 28, 55
Productivity, 111, 113, 138;
factor, 17, 28-29;
growth of, 29, 47, 51, 193
Profit, 64, 116, 127, 134;
net social profitability, 167, 182
Pro-natal, 8

Qin-Shi-Huang-di, 10
Quotas, 95, 127, 129-130, 134, 167, 169, 197;
procurement, 128

Ralston, K.L., 20
Rationing, 93, 137, 192, 199
Reforms, agricultural, 26-27;
industrial, 31, 37;
institutional, 196;
land, 9, 11-12, 76, 112-113, 126;
price, 37;
rural, 26-27, 29, 31, 37
Resource allocation,
agricultural, 122;
intersectoral, 121
Resources, natural, 12, 156, 166
Responsibility, 27, 127
Revolution, Cultural, 52, 193
Rice, 19, 51, 76-77, 82, 100, 158, 167, 173, 176, 196
Ruttan, Vernon W., 47, 49, 55, 95

Savings, 15, 189
School, 7
Security,
national, 164;
social, 93
Self-sufficiency, 6, 105, 126-127
Shi, Ru, 17, 145
Sicular, Terry, 125
Silk, 149, 158
Singapore, 176
Socialism, 126, 137
Sorghum, 84, 103
Soviet Union, 2, 150, 157
Soybeans, 20, 146, 148, 165, 178, 200
Specialization, 19-20;
areas of production, 12
Stability,
political, 93
Stagnation, 21, 50, 121, 154
Stalin, Josef, 50, 128
State, 6, 13, 94, 126-127, 130;
Statistical Bureau, 27, 30-31, 56, 195
Stone, Bruce, 51, 59
Strategy, 10-11, 20, 50, 99, 117, 147, 155, 157, 200;
development, 50, 147, 155, 188, 200
Subsidies, 17, 30, 32, 94, 105, 147, 197
Substitution,
factor, 55
Sugar, 150, 159, 165, 168-169, 174
Supply, 77, 96, 135, 155-156, 158, 189, 190-192, 194-198;
curve, 95, 129-130;
surplus, 106

Taoism, 10
Tariffs, 111, 166-167, 174
Taxation, 94;
burden, 95;
in-kind, 95, 100, 128;
product, 95, 128-130
Tea, 81, 150, 156
Teams,
leaders of, 36, 54;
production, 14, 36, 51-52, 56

Technology,
capital-using, 163;
improved, 47, 80, 106, 108;
labor-using, 163;
land-saving, 47, 54-56;
mechanical, 121
Tenant, 112;
landlord, 76
Tenure, 109;
uncertainty of, 19, 94, 196
Thiesenhusen, William C., 138
Tian-an-men, 10
Tian, Wei-min, 125, 132
Tien, Jiun-mei, 138
Timmer, C.P., 103
Totalitarian, 94
Township, 36, 75;
industries, 36
Tractors, 14, 56, 58-59, 65
Trade, 59, 111, 126, 147, 149-150, 156-157, 159, 168, 179, 200;
agricultural, 111, 145, 147-150, 156, 158, 166, 175-176;
agricultural export, 145, 150, 156, 158;
agricultural import, 148, 166, 168;
balance of, 111, 175;
competition in, 145;
international, 9, 20, 146, 166;
liberalization, 147, 164, 167, 179;
nonagricultural, 96;
surplus of, 111;
terms of, 93, 99-100, 120, 161
Trade barriers,
nontariff, 111, 169;
tariff, 111, 166-167
Trade fairs,
rural, 27
Tuan, Francis C., 17, 145
Transportation, 20, 80, 140

Unemployment, 32, 137
United States, 2, 6-7, 20, 68, 74, 175-176
Urbanization, 106, 139
United States Department of
Agriculture (USDA), 105, 167
Vegetables, 35, 82-83
Village, 30;
leaders, 32-33;
officials, 34-35
Volume,
procurement, 95
Wages, 32, 106
Wang, Guangsen, 93, 95-96
Wang, Y.T., 106
Webb, Shwu-eng H., 17, 19-20
Wen, Guanzhong James, 28
Whalley, John, 28, 52
Wheat, 19-20, 51, 150, 158
Wong, Lung-fai, 16-17
Workers,
contract, 32, 36-37;
hired, 35
World Bank, 71, 106, 189
Wu, Carson Kung-Hsien, 94, 97

Yeh, Shin-hsing, 138, 148
Yields, 50, 196;
high-yielding varieties, 10, 86
Yusuf, S., 16-17, 59

Zhao, Huazhou, 34
Zhong, F., 20
Zhu, Daohua, 148
Zhu, Li-jing, 52
Zhu, Rong, 51

For Product Safety Concerns and Information please contact our EU representative GPSR@taylorandfrancis.com
Taylor & Francis Verlag GmbH, Kaufingerstraße 24, 80331 München, Germany

www.ingramcontent.com/pod-product-compliance
Ingram Content Group UK Ltd.
Pitfield, Milton Keynes, MK11 3LW, UK
UKHW022203190926
475702UK00003B/173
* 9 7 8 0 3 6 7 2 8 6 2 8 6 *